Mathematical Conversations within the Practice of Mathematics

Studies in the Postmodern Theory of Education

Joe L. Kincheloe and Shirley R. Steinberg
General Editors

Vol. 129

PETER LANG
New York • Washington, D.C./Baltimore • Bern
Frankfurt am Main • Berlin • Brussels • Vienna • Oxford

Lynn M. Gordon Calvert

Mathematical Conversations within the Practice of Mathematics

PETER LANG
New York • Washington, D.C./Baltimore • Bern
Frankfurt am Main • Berlin • Brussels • Vienna • Oxford

Library of Congress Cataloging-in-Publication Data

Gordon Calvert, Lynn M.
Mathematical conversations within the practice
of mathematics / Lynn M. Gordon Calvert.
p. cm. — (Counterpoints; vol. 129)
Includes bibliographical references and index.
1. Mathematics—Study and teaching. I. Title.
II. Counterpoints (New York, N.Y.); v. 129.
QA11.G67   510'.71—dc21   99-049681
ISBN 0-8204-4598-3
ISSN 1058-1634

Die Deutsche Bibliothek-CIP-Einheitsaufnahme

Gordon Calvert, Lynn M.:
Mathematical conversations within the practice
of mathematics / Lynn M. Gordon Calvert.
−New York; Washington, D.C./Baltimore; Bern;
Frankfurt am Main; Berlin; Brussels; Vienna; Oxford: Lang.
(Counterpoints; Vol. 129)
ISBN 0-8204-4598-3

Cover design by Joni Holst

© 2001 Peter Lang Publishing, Inc., New York

All rights reserved.
Reprint or reproduction, even partially, in all forms such as microfilm,
xerography, microfiche, microcard, and offset strictly prohibited.

for Dennis ...

Table of Contents

Acknowledgments		ix
Prelude:	Conversation in Motion	xi
Chapter 1:	Introduction	1
Chapter 2:	Mathematical Inquiry Through Argumentation	13
Chapter 3:	The Practice of Mathematics	25
Chapter 4:	Creating Space for Mathematical Conversations	45
Chapter 5:	Explanations in the Process of Understanding	57
Interlude:	Improvisational Acts	87
Chapter 6:	Relationships Within Mathematical Conversations	89
Chapter 7:	Lingering in a Mathematical Space	113
Chapter 8:	Re-Turning to Features of Mathematical Conversation	131
Appendix A:	The Research	151
Appendix B:	Interactions Not Consistent with Features of Mathematical Conversation	157
References		161
Index		171

Acknowledgments

Each word herein is filled with the guidance and compassion of those with whom I have dwelled in the bringing forth of this work.

Thank you,
    for expanding my cognitive domain;
            Tom Kieren, Elaine Simmt
            Al Olson, David Blades, Marg Iveson
            Erna Yackel, Julia Ellis
            Past and present residents of 948

    for creating space for me to breathe and grieve;
            Robert Jackson

    for opening your arms to me and for the many times you carried me.
            Gail, Diane, Debbie
            Terry, Joanne, my parents
            Paula, Scott, Rhonda
            My slowpitch friends

Prelude: Conversation in Motion

We face one another
        mirrored images in motion
Palms up
        feigning surrender
I reach out my arms
        stretching the seams of my
        boundaried box

She follows (my lead)
        slightly behind
a timed delay of lightening
                thunder
an annoying lag between picture
                sound

Willing her to take the burden of control
I pass leadership to her in frozen silence

She begins
        a slow controlled sideways lunge
I watch her eyes
        windows through her body
trying to imitate, accommodate,
                anticipate
We continue halt
        ingly
    lead
        freeze

Passing leadership like pepper
                from one to an
                                other

Then
without knowing how or when I suddenly
realize through our movements
that I no longer know who is leading
                        who is following

Neither of us
      both of us
            the movement itself
No start
    stop
Just a continual flow of movement
a common rhythm
one movement flowing into the next
one movement an echo of a past
      an anticipation of a future
A conversation in motion.

Chapter 1

Introduction

I have always loved math—or what I experienced as school mathematics. I liked doing fifty questions out of the math text. Even though there was generally little difference between one question and the next, each question, each problem was a game to be played—a game that I usually won.

One of my favorite memories of mathematics is of doing modules in elementary school. Each module was a separate booklet which we had to complete independently and then take to our teacher, Mrs. Mayfield.[1] She would correct it with her red pen, and if she was satisfied with our work we would sit at the back table and write a timed test. If we completed it within the time limit with less than the set number of errors we would get a star beside our name on the class list and a brand new booklet. Completing the booklets fed into my competitive spirit. I wanted to get as many stars by my name as possible and to do so I needed silence with no disruptions from either my peers or the teacher.

Unlike many of my peers, I was fairly successful learning mathematics via the 'chalk and talk' approach. What was written on the board and 'transmitted' to us through lecture almost always made perfect sense to me. When it didn't, I rarely asked questions; instead, I would work it out myself—independent of help from peers, teachers or family. My desire to work independently continued throughout my schooling. I felt a deep sense of satisfaction and joy from working through a problem on my own and getting the right answer. For me, working with others while doing mathematical activities would have been annoying as well as intimidating. If I had been asked to create, explore, and extend mathematics as opposed to finding a clear and predictable answer, I would likely have become frustrated and resentful. The long straight rows, absolute silence, and clear direction and purpose was, at least for me, the ideal mathematics classroom.

I place these experiences up front because in many ways they completely contradict what I present in this exploration of mathematical conversations. Although it will take the whole of this book to describe the features of mathematical conversation as a discourse and its implications for the practice of school mathematics, I could perhaps begin by stating what this work is not. It is not research on conversations as mathematical "talk." It does not reach the conclusion that achievement improves or more mathematical content is learned through interaction. And although it has implications for teaching and classroom activity, the work does not develop a prescriptive model for teaching or promote particular student activities. Rather, this work describes and interprets mathematical conversations within the whole of doing and coming to know mathematics with others.

Mathematics instruction has a long history in classrooms as teacher-directed communication with sequential and discrete skill-oriented practice. Recent North American reforms in mathematics education, led by the National Council of Teachers of Mathematics (NCTM), have encouraged teachers to alter discourse and practice in the mathematics classroom. Reforms suggest shifting the authority of knowledge from the teacher to the classroom as a community by incorporating an argumentative discourse situated within a problem-solving practice. These reforms have arisen from perspectives in the philosophy of mathematics and from present theories of cognition (see chapters 2 and 3). This work explores the potential for a discourse of mathematical conversation by building on these present reforms in mathematics education, by addressing recent developments in the philosophy of mathematics, and by choosing an *enactivist* framework for knowing and language which places an emphasis on cognition as it occurs in culturally and historically situated interactions with others and the environment (see chapter 4). *Mathematical Conversations Within the Practice of Mathematics* attempts to provide an alternative vision of discourse and practice for mathematics education.

The words *discourse* and *practice* have been chosen specifically to communicate the all-at-once quality of knowing, language, interaction, and mathematical activity within the classroom. The usual definition of discourse is "the communication of thought by speech" (O.E.D. 1992). Rather than a simple focus on the words spoken, however, the use of discourse here refers to the original sense of the word. The etymology of 'discourse' stems from the Latin word *discurrere* in which *currere* means "to run" as in, to run a course. Yet, rather than envisioning a predetermined course to be run, a discourse of conversation may be thought of as "the running of the course" (Pinar and Grumet 1976) or as the act of "laying down a path in walking" (Varela 1987). However, this course or path is not random or fully open. Discourses proscribe

> what can be said, and thought, but also about who can speak, when, where and with what authority. Discourses embody meaning and social relationship, they constitute both subjectivity and power relations. . . . Thus, discourses construct certain possibilities for thought. They order and combine words in particular ways and exclude or displace other combinations. (Ball 1990, 17)

Issues of power, the patterned flow of actions, and the constraints on ways of knowing and thinking are embedded in the discourse enacted in that moment. Discourse, then, is not used simply to communicate nor is it used to construct a *view* of reality as though a 'real' world exists and is visible outside of us. Through our language and discourse we *bring forth* a reality (Maturana and Varela 1987). Through a mathematical discourse we bring into being mathematical objects, concepts, theories, and proofs to make sense of our experiences.

The word *practice* has two meanings. Our experience with it in school mathematics has been connected to the notion of "practicing mathematics"—that is, the repeated performance or execution of a skill to gain proficiency. If the sequence of performance becomes totally known to the individual, which does not necessarily require a depth of understanding, it becomes habit or "ritual" (Bateson 1979). The pattern of action becomes entrenched and the same path is followed each time. Once a ritual is formed it is difficult for a person to imagine a new path, to bring forth an alternative possibility, a different reality.

The "practice of mathematics," on the other hand, describes the profession of mathematics: What are the legitimate activities of a person who claims to be a mathematician? Unfortunately, in the past we have presented to our students the practice of mathematics as "practicing mathematics." Mathematical techniques and algorithms, the tools of mathematical practice, have simply been practiced repeatedly with the hope that they become rituals—rituals used for some potential future purpose.

Throughout this book, mathematical practice and the practice of mathematics are used interchangeably to mean the socially agreed upon activities of a person engaged in the doing and creating of mathematics, whether that person is a professional or amateur mathematician. The practice of mathematics defines the nature of, the tools for, and the means for expressing the play, work, and action within mathematics. Asking, "What is mathematical practice?" is synonymous here with the question, "What is it like to engage in mathematics?"

I present to the reader mathematical conversations as an alternative image of mathematical practice and discourse. I do so not with the intention of insisting how life should be or must be in the mathematics classroom, but rather with the intention of questioning assumptions we may hold about present practices and discourses. Raising concerns about mathematical practice and discourse provides opportunities to open ourselves and broaden our horizons to alternative possibilities (Gadamer 1989). If one is to accept mathematical conversations as a means for engaging in legitimate forms of mathematical practice, an alternative image of both practice and discourse is required.

Re-Searching Mathematical Conversations

> D: Daddy, why do things have outlines?
> F: Do they? I don't know. What sort of things do you mean?
> D: I mean when I draw things, why do they have outlines?
> F: Well, what about other sorts of things—a flock of sheep? Or a conversation? Do they have outlines?
> D: Don't be silly. I can't draw a conversation. (Bateson 1972, 27)

Outlining the Work

My areas of interest in mathematics education have been in the fields of communication, language, interaction, cognition, creativity, intuition, and so on. A few years ago I began playing with these words and several related ideas in a web diagram. The page became filled with words, phrases and connecting branches. I then stopped, looked at the whole page at once, and wrote "genuine conversations" in the middle of the sheet. As I stared at the words I drew a dark heavy circle around them. I wondered, "Is it possible to have a genuine mathematical conversation?" Although I assumed conversations did occur, I wondered, "What might mathematical conversations be like, and do they serve any purpose?"

> To ask a question means to bring into the open. The openness of what is in question consists in the fact that the answer is not settled. (Gadamer 1989, 363)

To me, a genuine conversation wasn't an exchange in which one person tried to "teach" the other person something by leading him or her through a series of steps. It also wasn't an argument between people debating over who was right and who was wrong when their mathematical answers differed. I saw a mathematical conversation as something where both persons, together, were trying to come to an understanding of some mathematical phenomenon. While I thought mathematical conversation might be a viable means through which to learn mathematics, I questioned its potential given my own experiences as a student and wondered whether it would give rise to legitimate or appropriate mathematical activity.

I brought these questions and thoughts to a research project I was involved in in which pairs of undergraduate and high school students were given mathematical prompts in clinical settings.[2] The paired students were asked to work together on open-ended investigations while researchers observed their mathematical actions and understandings. I began to view every pair's interaction with the question in mind: "Are these people engaged in a mathematical conversation?" Although sometimes my answer was no, particularly when the students chose not to interact with one another at all, many times it "felt" like they were, at least for some portion of their investigation. When such occasions occurred I observed more closely with the following question in mind: "What are the 'qualities' or 'features' of their conversation and what is the nature of their mathematical practice?"

Over the next few years I had the opportunity to observe and interact with students who appeared to be engaging in mathematical conversations. Some of these students were subjects in research projects like that described above; some were undergraduate students that I asked to help me with my research; and another site was an extracurricular program involving parents and children engaged in mathematics together.[3]

Gadamer (1989) writes that "understanding begins. . .when something addresses us" (299). Whenever I found myself in situations where students engaged in open-ended mathematical activities—whether they were students from the sites mentioned above, or from my junior high math club or grade one math lab, or within my elementary methods classes—questions concerning mathematical conversations constantly *addressed me*. The collection of data for this work was and is not complete. It is necessarily ongoing because I cannot help but "research" the topic whenever and wherever the environment is conducive to mathematical conversation. As I was writing my findings, I was also "finding" in my writing. The research question, the data collection, the interpretations of data, the writing of interpretations. . .these research phases occurred *all-at-once*.

All-At-Once Methodology

Rather than a purely inductive or deductive approach, the methodology and theory presented in this work occurred "all-at-once" through the interaction of a large number of sources and research activities always "occurring in the now." It was an ongoing process with multiple overlapping layers of data sources and multiple interacting interpretive activities. A wide range of data artifacts was collected through video- and audiotapes, transcripts, field notes, interpretive diagrams and research dialogue journals. The methodology was not systematic and dutiful to pre-established procedures and methods. The artifacts of the research were continually collected as new research participants became involved, and previous participants' work was reinterpreted in light of new data. Interpretive activities were not restricted to particular techniques of data analysis. Therefore, the work was not simply a matter of "writing up" the findings. The very act of writing was an important interpretive process.

It is also important to recognize that the interpretations reported here are not mine alone. Ideas, theories, and interpretations are not autonomous creations. As Bakhtin (1981) writes, "The word in language is half someone else's" (293). Readings from educational and social science research and philosophy also interacted with the data collected. Conversations and the sharing of writing within my community of colleagues frequently allowed me to think differently about the data, descriptions, and interpretations made. A presentation of this work and the subsequent discussion was also an opportunity to re-visit interpretations made to that moment and expand my horizon of understanding further. These interactive events folded into my thinking and subsequently altered the way I read, wrote, and observed future research activities.

The reporting of this data cannot be thought of as presenting a correct view of what "is" or what "was," for no such view is possible (Shotter 1993). Every time I turned to address or respond to a specific question or concern within the research, I simultaneously turned away from and closed off other questions and aspects of the data (Gadamer 1989). My descriptions and interpretations, then, are necessarily incomplete and open to further interpretation. Their incompleteness is not because they stand as only a partially correct match to a universally objective truth, but because they "initiate and guide a search for meanings among a spectrum of possible meanings" (Bruner 1986, 25). Interpretation and explanation allow us to enter into the creation of meaning and are purposeful within their incompleteness to shape expectations and provide meaning and coherence for experiences related to mathematical interactions. It is expected that a reader will expand upon these explanations further so that an even broader understanding of the experiences may be shared. The incompleteness of descriptions, explanations, and interpretations is not a limitation of this work or of our selves as humans, but is a condition and feature of our human existence in general (Maturana 1998). Although the set of interpretations developed in this work will attempt to make a number of generalizations, these generalizations are not universal truths standing for always and forever, but attempt to be temporally coherent and meaningful in their plausible, possible, and incomplete state.

Drawing Conversations
Returning to Bateson's quotation about the impossibility of drawing a conversation, I ask, "How is it possible to capture conversation on a page?" As researchers we frequently find ourselves trying to dissect conversation—slicing a scalpel through the embodied words; leaving them lifeless and pinned down on the page.[4] In our attempts at capturing and controlling conversation, we may grab hold of the words spoken, but we let the living voices slip through our fingers. To understand the conversations of others does not mean magnifying the particles and particulars of words or phrases. Instead, there is a need to pay attention to the rhythms and textures of conversation, for the meaning is inseparable from the sound, the tone,

> the silences . . .
>   the rhythm of bodies, the resonance of words.

Words, emotions, nonverbal expressions, action, and the trail of history brought forth in the moment, all flow together to form the meanings in conversation. Conversations are, at all times, incomplete. They possess a degree of ambiguity and vagueness that prevents their capture and full description. Conversations remain at play, defying closure; they are never finalized or contained. They do not wholly exist.

Re-presenting mathematical conversations as they occur all-at-once and in the now presents difficulties.[5] Simply including a transcription of the words spoken would contradict the image of conversation as culturally and historically situated utterances and actions. Instead, I have tried to present the conversations in context through the use of narrative; yet, with the recognition that this form will also be incomplete, for in writing the narrative, I "select" what aspects of the conversation to highlight and necessarily ignore others. My selections are based on the conversational features that surfaced and addressed me, and the questions, issues, and concerns that arose as I observed the conversation. Selections were also based on questions, concerns, and conversational features that I held and also those brought to my attention by the participants and by colleagues with whom I interacted. Through my selections and narrative portrayals, I attempt to draw the reader's attention to particular features and aspects of the conversation that may not have appeared if merely the words were printed, or may have gone unnoticed if a videotape was presented of their words and actions. It is expected that readers will not simply "receive" the narratives and the interpretations of them as is, but will interact with the text to construct their own images and interpretations of the events that took place and expand on what is offered here.

Overview

The chapters of this book are an exploration of mathematical conversations and their place in mathematical practice. The intention of the work is, first, to provide a basis for understanding mathematical conversations within the practice of mathematics; second, to explore mathematical conversations through the use of illustrative examples; and, third, to summarize the features of discourse and practice within mathematical conversations.

Basis of Mathematical Conversations

In chapter 2, "Mathematical Inquiry Through Argumentation," the role of argument within the practice of mathematics is discussed by examining the impact that recent philosophers in mathematics such as Polya and Lakatos have had on our approach to mathematics instruction. Reforms in mathematics education have utilized information in the philosophy of mathematical practice to envision and implement an argumentative discourse within an environment of problem solving.

Chapter 3, "The Practice of Mathematics," addresses assumptions underlying a problem-solving curriculum, such as the view that problem solving is necessary for living in a technological world and the perspective that problem solving is synonymous with intelligence. In chapters 2 and 3 the theoretical, epistemological and pedagogical assumptions underpinning an argumentative discourse and a problem-solving practice are questioned.

Chapter 4, "Creating Space for Mathematical Conversations," presents the epistemological and theoretical basis for mathematical conversations. This chapter describes an enactivist framework for learning and for living using the work of biologists Maturana and Varela along with the study of hermeneutics by Gadamer. The theoretical framework emphasizes the ongoing interactions between persons and between persons and their constituted world. This chapter examines how mathematical conversations are currently described in the mathematics education literature and attempts to provide a broader understanding of conversation using literature in philosophy and language.

Illustrative Examples of Mathematical Conversations
Chapters 5 to 7 present illustrative examples of mathematical conversation by featuring the work of three pairs of students from three research sites. Data for the chapters were selected to highlight a variety of overlapping conversational features.

Chapter 5, "Explanations in the Process of Understanding," is the first of the three illustrative examples. The importance of student explanations has been frequently highlighted in recent mathematics education research. This chapter examines the nature of mathematical explanations as they occur in a mathematical conversation by two university students. It draws attention to the formulation of concerns, how explanations are offered, and what is accepted as a mathematical explanation in conversation. Arising from this illustrative example is the broader question, "What is accepted as a mathematical explanation within the discipline of mathematics?"

Chapter 6, "Relationships Within Mathematical Conversations," explores the interactive dimensions of mathematical conversations; that is, the formation, maintenance, and reparation of relationships between the persons engaged in conversation and between the persons and their mathematical environment. Two university students engaged in mathematical experimentation provide an illustration of how relationships with others and with a mathematical phenomenon investigated are maintained and repaired. This chapter explores the nature of these two students' play and highlights the importance of creating a space for the other cognitively, physically, and emotionally in mathematical conversations.

Chapter 7, "Lingering in a Mathematical Space," presents the third and final illustrative example. A description of mathematical activity between a mother and her son is offered to illustrate the "lingering" nature of their mathematical engagement. Rather than following a pre-established course or directing their activities towards a predetermined goal, their mathematical conversation proceeded as the laying down of a path which was shaped interactively by the subject matter and the phenomenological histories that each person brought to the conversation. This chapter questions the role of the teacher in a mathematical conversation, not as one who leads, or even facilitates, but as one who engages in conversation with his or her students. It also examines the definitions for intelligent action within the practice of mathematics. Accepting lingering as intelligent mathematical activity requires that one "play the game" of mathematics differently.

Consequences and Questions
Chapter 8, "Re-Turning to Features of Mathematical Conversation," revisits and summarizes the theoretical features of mathematical conversations in terms of the interactive relationships between persons and between persons and the mathematical world they bring forth, and in terms of the curriculum that is lived as a result of these interactions. The possibilities, potential difficulties and consequences for implementing mathematical conversations in the classroom are addressed in this final chapter.

Invitation to the Reader

This work attempts to articulate the living practice of persons engaged in mathematical conversations. It does not attempt to shape these practices into an objective ideal or method. Instead, this work "offers an understanding of human life, not as an object with properties, but as constituted by a lifelong movement of interpretation and re-interpretation, of reading and re-reading that is always caught in the moment of arriving" (Jardine and Field 1996, 256). The understanding that is sought through interpretation is a response to the question, "What is it like to engage in mathematical conversation?"

Mathematical conversation within the practice of mathematics attempts to build on recent reforms in mathematics education, rather than reject them. It attempts to push the present boundaries of mathematical practice and interaction, rather than draw out a new field of play. My purpose is not to convince readers to change their beliefs about mathematical discourse and practice; rather, readers are invited to enter into conversation with the text, to question their own assumptions and expectations regarding discourse and practice, and to question the underlying assumptions presented here about the nature of mathematics as well as mathematics teaching and learning. It is hoped that by doing so, we can continue the conversation about mathematical discourse and practice as it occurs in our classrooms.

Notes

1. The name has been changed to protect anonymity.
2. The Mathematical Understanding Project led by Dr. Tom Kieren, University of Alberta.
3. The data in this work uses one pair of participants from each of these three sites. A more detailed description of the sites and the expected activities occurs in Appendix A, as well as in chapters 5, 6, and 7, where the interactions are illustrated.
4. Research methods that involve "slicing" include the act of searching for repeated words, or themes from words; contrasting A's speech pattern with B's; classifying speech acts; and a variety of other discourse analysis methods.
5. "Re-present" throughout this work implies reconstructions or reflections of previous experiences rather than conceptual pictures or replications of a seemingly independent objective world (von Glasersfeld 1987).

Chapter 2

Mathematical Inquiry Through Argumentation

The Philosophy of Mathematical Practice

There is a shroud of mystery over what it is that mathematicians do. For many people the image that may spring to mind is of a mathematician standing at the chalkboard scribbling a series of indecipherable symbols, numbers, and Greek letters. The mathematician rarely hesitates between lines and soon the whole board is filled with white dust. Once completed he steps back, smiles at the beauty of the 250-year-old proof before him, and says, "This proof is intuitively obvious isn't it?"[1] The only exposure many people have to mathematicians is through university courses. While this image may reflect a person's experience with mathematics and mathematicians' instruction, a mathematician's research activities are often completely unknown. Even within the field of mathematics, a mathematician's practice is often narrowly described as verifying conjectures through rigorous proof. Only recently have there been attempts to expand this description to include "such quasi-empirical topics as discovery and communication, informal proofs, errors, explanations, history or cultures, computers or psychology" (Tymoczko 1986, 127). Our reason, as teachers, for becoming aware of the practice of mathematics is not because our goal in education is to prepare our students to be mathematicians by mimicking this practice. Rather, understanding how new mathematical knowledge is generated and accepted by the discipline is essential for understanding what it means to know and acquire mathematical knowledge and for becoming aware of how this knowing and acquisition may take place in the classroom (Putnam, Lampert and Peterson 1990).

The mathematician and educator George Polya noted that it was not enough to understand how theorems are proved; one must also understand how conjectures are discovered in the first place. Polya (1973) described a reasoning process he called "heuristics" in problem solving. "Heuristics endeavors to understand the process of solving problems, especially the mental operations typically useful in this process" (130). Heuristics is a study of the "methods and rules of discovery and invention" (112). Polya described what is now a classic and clichéd four-step procedure for completing problems: "understanding the problem," "devising a plan," "carrying out the plan," and "looking back." Within this process Polya identified a number of strategies involved, such as "make a chart," "work backwards," "guess and check," and "look for a pattern." Recent mathematics education research, instructional and assessment resources, as well as mathematics curricula draw heavily on Polya's work in search of a 'method' for teaching problem solving.[2]

Polya's interest in describing the problem-solving process was in part a reaction against formalism, a foundationalist philosophy of mathematics which perceives mathematics as a formal system of axioms, rules and procedures with no tangible meaning in and of themselves. Polya insisted that many features of mathematical practice were shared with science, such as induction, plausible reasoning, guessing and analogy. While a formalist perspective emphasizes how theorems are formally proved, Polya (1954) suggested that this form of reasoning, which he called "demonstrative reasoning" was of secondary importance. It did not yield new knowledge about the world around us. It was "safe, beyond controversy, and final" (v). Demonstrative reasoning, step-by-step, linear and error-free, as found in formal proofs, is often the image of mathematics presented in schools and at universities. As a result, students are often left with the impression that mathematics arises from one's mind in a finished and polished form; errors, tangents and uncertainty are thought not to occur if one is a knowledgeable mathematician.

Polya (1954) insisted that "mathematics in the making resembles any other human knowledge in the making" (vi). Prior to demonstrative reasoning, Polya suggested that "plausible reasoning" was at work. That is, reasoning in the form of conjectures or "guessing" through generalization, specialization and analogy. Here, he suggested, was where new learning occurred—in its controversial and provisional state.

> You have to guess a mathematical theorem before you prove it; you have to guess the idea of the proof before you carry through the details. You have to combine observations and follow analogies; you have to try and try again. The result of the mathematician's creative work is demonstrative reasoning, a proof; but the proof is discovered by plausible reasoning, by guessing. If the learning of mathematics

reflects to any degree the invention of mathematics, it must have a place for guessing, for plausible inference. (vi)

Discussions and descriptions of mathematics in the past often portrayed mathematical practice as a logical and linear process of reasoning leading to theorems. Polya's work draws attention to plausible reasoning leading to conjectures and the difficulty of describing such unformulated, unformulaic, and informal reasoning.

Imre Lakatos (1976) wrote of the relationship between conjecture and proof in his book *Proofs and Refutations*. His study detailed case histories of how particular mathematical problems in Euclidean geometry have come to be formulated and resolved. Lakatos demonstrated that problems often evolved through a chain of reformulations, counterexamples, and partial proofs (Barrow 1992). Lakatos' aim was to return fallibility and humanness to the practice of mathematics by proposing a mathematics methodology that was "quasi-empirical" and similar to scientific methods of research. Lakatos offered his work as a criticism of formalism and dogmatic philosophies that were prevalent at that time. He argued that defining mathematics as simply a formal system left no place for the "logic of discovery" and left out the cultural and historical development of mathematics; that is, the humanistic component behind mathematics. Like Polya, Lakatos (1976) suggested that "informal, quasi-empirical, mathematics does not grow through a monotonous increase of the number of indubitably established theorems but through the incessant improvement of guesses by speculation and criticism, by the logic of proofs and refutations" (5). He described a quasi-empirical method as a dialectic in which hypotheses were generated and "weeded out by severe criticism" (Lakatos 1986, 34). He believed that proof was never final or certain, as the assumptions on which the proof was founded were always open to reexamination by the mathematics community. Lakatos' work provided a philosophy of mathematical practice in which mathematics was viewed as a social phenomenon; proofs were accepted not because they followed a formalist system of rules and axioms, but rather because they adhered to a discourse of argument within a mathematical community.

Argumentation as a Form of Inquiry

Argumentation in Mathematics Education

Our current reform image of mathematical discourse has been derived from references to disciplinary practices as described by Polya and Lakatos. Research in mathematical discourse has focused primarily on dialectic or argumentative approaches as Lakatos described (e.g., Lampert 1990; Lampert, Rittenhouse, and Crumbaugh 1996). Attention to discourse and language in mathematics education is consistent with a current sociocultural emphasis in educational research which attends to how discourse and language are cultural and cognitive mediators of learning (Hicks 1996a). Hicks (Hicks 1996b) states that

> Educational standards across domains now frequently invoke the notion that students need to be able to communicate in ways that reflect authentic disciplinary practices. Children need to be able to 'talk science' and 'talk math' in order to participate fully in the reasoning processes and social practices that characterize those disciplines. (50–51)

Current research in mathematics education has paid particular attention to discourse. There is a strong desire to move from a "teaching as telling" model to a model of the "classroom as a community of learners." An outcome of this work is evident in the "students' role in discourse" as stated in the *Professional Standards for Teaching Mathematics* (NCTM 1991).

> The teacher of mathematics should promote classroom discourse in which students—
> - listen to, respond to, and question the teacher and one another;
> - use a variety of tools to reason, make connections, solve problems, and communicate;
> - initiate problems and questions;
> - make conjectures and present solutions;
> - explore examples and counterexamples to investigate a conjecture;
> - try to convince themselves and one another of the validity of particular representations, solutions, conjectures, and answers;
> - rely on mathematical evidence and argument to determine validity. (45)

The influence of Polya and Lakatos and their descriptions of mathematical practice are highly visible in these statements. Making conjectures, exploring examples and counterexamples, as well as initiating problems and questions emphasize the "logic of discovery." The social nature of mathematics is apparent in these statements, as students are expected to present convincing arguments as a process for validating their answers. The discourse described above drew upon mathematics education research in classroom discourse at that time and it is still reflective of current research today. The statements reveal the extent to which mathematics as *problem solving* and as *argument* underlie our current reform movement in mathematics education. Recent research in mathematical discourse continues to reference the use of argumentative approaches for mathematics learning (e.g., Cobb, Boufi, McClain, and Whitenack 1997; Krummheuer 1995; Lampert 1990; Lampert, Rittenhouse, and Crumbaugh 1996; Yackel and Cobb 1996).

Lampert (1990) explicitly draws on the work of Lakatos and Polya and their complementary descriptions of mathematics as a discipline to establish a *culture of disagreement* in the classroom. This discourse attempts to support mathematical learning through

> a process of "conscious guessing" about relationships among quantities and shapes, with proof following a "zig-zag" path starting from conjectures and moving to the examination of premises through the use of counterexamples or "refutations". (30)

In a later work, Lampert, Rittenhouse and Crumbaugh (1996) describe a classroom in which Lampert explicitly teaches the discourse of disagreement to her students through a set of norms. Students are to use phrases such as "I agree with Andy because" or "I disagree with Janaya because" emphasizing the reasoning and justification of such agreement or disagreement.

Krummheuer (1995) describes a similar discourse in his "ethnography of argumentation." Drawing on the tradition of rhetoric traced back to Aristotle, he defines argumentation "as a process that is accomplished by a single person confronted with an audience that is to be convinced" (231). "Its aim is to convince oneself as well as other participants of the property of one's own reasoning and to win over the other participants to this special kind of 'rational enterprise'" (247).

The image of mathematical activity through the discourse of disagreement and argumentation has done much to provide an image of mathematics as a human endeavor with truths founded on and socially negotiated through argument. By appropriating this discourse, participants, including both teacher and students, attempt to situationally define what counts as knowledge or knowing within the mathematics classroom. Rather than a focus on finding answers, the intent is to focus on justification and explanation for supporting or rejecting strategies and solutions offered by students within the classroom. The responsibility for determining correct or acceptable answers shifts from teachers and textbooks to the classroom members as a community of learners.

An argumentative discourse in the mathematics classroom is well supported by the literature in the philosophy of mathematics and in contemporary learning theories. Argumentation also has appeal beyond mathematics education.

Argumentation Nonspecific to Mathematics

Argumentation, as a mode of discourse, is not specific to mathematics. In education, critical pedagogy and critical theory stemming from Paulo Freire's work are also forms of argumentative discourse that move the determination of right and wrong from an authority to the community. Even more broadly, in Billig's book, *Arguing and Thinking* (1987), argumentation is described as a general form of inquiry used throughout the physical and social sciences. Argumentation as a mode of discourse is related epistemologically to critical rationalism, an anti-positivist perspective for the growth of knowledge. Popper, a strong advocate of critical rationalism (and also a major influence in Lakatos' work), argued that the method of conjecture and refutation was at the heart of *all* processes of knowledge growth. Rationalism, as described by Popper, is a readiness to listen to critical arguments. "It is fundamentally an attitude of admitting that 'I may be wrong and you may be right and by an effort, we may get nearer to the truth'" (Popper 1945, 225).

Argumentation has a stronghold in our society, particularly among the academic elite. It is the form of discourse used in published papers, at conference presentations, within graduate seminars, and at formal debates. Yet it is very difficult for many people, both children and adults, to fully participate in this form of discourse.

The Difficult Nature of Argumentation

Argument is War

In their provocative work on the *Metaphors We Live By*, Lakoff and Johnson (1980) describe the metaphor, "Argument is War."

> We don't just talk about arguments in terms of war. We can actually win or lose arguments. We see the person we are arguing with as an opponent. We attack his positions and we defend our own. We gain and lose ground. We plan and use strategies. If we find a position indefensible, we can abandon it and take a new line of attack. (4)

The underlying goal in many arguments is to determine a winner and loser, often at the expense of seeking a broader understanding. An emphasis on winning may hinder one's ability to listen to the other's argument and simultaneously "caters too comfortably to our natural impulse to protect and retain the views we already hold" (Elbow 1986, 263).

When arguments are played for the goal of winning and losing, what roles do power and status play? Power is often assumed to be a negative force resulting in a "distortion and corruption of knowledge" (Ushers and Edwards 1994, 85). The aim of democratic and emancipatory discourses based on this assumption is to attempt to remove power and status from the playing field so that arguments are made and judged solely on the basis of reason and justification of reason; participants are not to be swayed by the authority of the person arguing. However, if we agree with Foucault (1980), power is omnipresent in all forms of interaction and discourse. Knowledge and power then, cannot be separated. From this perspective, a public forum, such as a whole class setting, is *not* an arena in which all participants can play on equal terms. Speaking time is a scarce resource and those who obtain the floor most readily are often those persons with the most influence through authority, power, status, or a gift of rhetoric—not necessarily those individuals with the most reasonable argument.

Educational Difficulties with Argument

Implementing an argumentative discourse in the classroom has been faced with criticism from sociological and epistemological points of view. One of the difficulties that a teacher faces when trying to implement such a discourse is the tension between the social norms students experience within their community of family and friends and those required by argumentation. Sociolinguistic research suggests that if classroom discourse is not consonant with community-based discourse practices, then children may experience difficulties (Hicks 1996a). As an example, after participating in a "discourse of disagreement" for most of a school term, Lampert asked her students to discuss their experiences of participating in the discourse (Lampert, Rittenhouse and Crumbaugh 1996). The children spoke predominantly of the negative emotions they felt engaging in this form of intellectual debate. Lampert et al. provided quotations offered primarily by Ellie, one of the children in the classroom. Below are some excerpts from the comments made:

> *Ellie:* I don't like reasoning because whenever you have a wrong answer people try so hard to prove you're wrong;
> *Saundra:* Yeah, I agree with Ellie because you know it can get sort of embarrassing at times, because like everybody else, like you say something and everybody will raise their hand and want to say something different or they all disagree with you. And it makes you sort of feel like you want to crawl into a hole and die.
> *Ellie:* When you do realize that you have the answer wrong they still want to prove it to you that it's wrong. . .you want to just crawl under your desk.

Ellie: Sometimes I don't like discussions because when you're trying to prove something it just turns into something else and you don't get to say what you think.

Ellie: It's fun to have an answer that's right, but if everybody is raising their hand and um, you're called on, you feel really bad if, if um, somebody really wanted that to um, give that answer. And sometimes people say, "Oh, you took my answer," and they don't talk to you for a while. (742–743)

Lampert, Rittenhouse and Crumbaugh summarize these quotes by saying, "It seems to be the case that these students experience 'people trying to prove you are wrong' as a personal assault, even in situations where the teacher insists that assertions be justified with mathematical evidence" (744). The feelings of being attacked are consistent with the metaphor, "argument is war" (Lakoff and Johnson 1980). Students felt defeated when wrong and victorious when right. The statements, "You don't get to say what you think" and "you feel really bad if . . . somebody really wanted . . . to . . . give that answer" are indicative of the role of power at work and the limited speaking time available within a whole-class setting. The primary focus of these students' comments was on their emotions of being right or wrong, winning or losing the argument. It is not, or at least it should not have been, surprising that these students were not able to become emotionally detached from the answers they gave. Intuitively we understand the difficulty, or perhaps impossibility, of asking students to separate themselves from their ideas.

Berkowitz, Oser and Althof (1987), studying peer conflict in adolescents, describe argumentation as an "ideal" form of discourse. Arguments focus on developing the best position, theory, or strategy, without concern for personal views, power, or status. Although ideal, the authors state that it is unlikely that many adolescents will reach this stage of dialogue. An added difficulty is that even if one person has the capacity to participate in this ideal discourse, it will not succeed without a partner or partners who are also capable of doing so. Creating an environment in which an argumentative discourse is possible and tolerable in a classroom or public setting appears to be a difficult task to achieve.

Argumentative discourse, similar in nature to a public debate, is assumed to be equitable and enlightening. However, the inequity of power and status has made argumentation and debate within academic arenas a target for criticism from feminist educators. Far from an ideal, this traditional form of rhetoric is considered patriarchal and is said to disadvantage women and nonwhite European males. For instance, Lewis and Simon

(1986) describe their participation in such a forum during a graduate seminar. "[The men] sparred, dueled and charged at each other like gladiators in a Roman arena. Yet their camaraderie intensified with each encounter. Throughout this exchange, the women were relegated to the position of spectators" (460–461). Lewis and Simon's image of the men's interaction is described as "war-like." While it appeared satisfying to the male students, it was described as threatening to and exclusionary of most of the female students.

Similarly, Ellsworth (1989) critiques discourse within the framework of critical pedagogy founded on Freire's work. Ellsworth states that argument as a form of critical discourse promotes the myth of the ideal rational individual: an individual that is capable of separating knowledge from emotions, personal interests and experiences. It is assumed that the removal of power, status, emotion, and personal interest are possible and necessary to allow "individuals to realise their inherent rationality, express themselves freely and develop themselves fully" (Ushers and Edwards 1994, 85). However, Ellsworth states that in her own experiences these classroom practices do not guard against the relations of domination they seek to overcome. Participants in her study agreed that "commitment to rational discussion. . .in a classroom setting was not enough to make that setting a safe space for speaking out and talking back" (316).

Although the previous section has presented a number of difficulties and criticisms regarding the implementation of an argumentative discourse into the classroom, it is still a highly regarded form of interaction in our society. One explanation is the deep-seated connection western culture has made between argumentation and knowledge acquisition. This is particularly true in the field of mathematics.

Separate Knowing and Paradigmatic Thought

Argumentation is similar to the ideal of "separate knowing" as described by Belenky, Clinchy, Goldberger, and Tarule (1986; Goldberger, Tarule, Clinchy, and Belenky 1996). The authors base this form of knowing on a method of inquiry described by Elbow (1973, 1986) as "the doubting game" or "methodological doubt." Derived from Descartes' "Method," a person searches for knowledge by trying systematically to *doubt* everything, even if it appears obvious. What cannot be doubted at the end of this process must be true. Playing the doubting game involves critically examining a piece of text or another person's ideas by consciously (or perhaps unconsciously if one does not realize he or she is playing this particular game) searching for flaws in reasoning: "a loophole, a factual error, a logical contradiction, the omission of contrary evidence" (Belenky, et al. 1986, 104).

The mathematical practice of proofs and refutations as described by Lakatos is framed within an inquiry based on methodological doubt. Once a conjecture has been made, a person attempts to find *counterexamples* that refute it and prove it false; hypotheses are generated and then weeded out by "severe criticism" (Lakatos 1986, 34). Similarly, Krummheuer (1995) suggests in his theory of argumentation, that the speaker's goal is to convince and win over his or her audience, and it is the audience's role to *doubt* the conclusions of the speaker. Argumentation as it has been described, has a strong correlation with separate knowing and the process of methodological doubt. "In separate knowing one takes an adversarial stance toward new ideas, even when the ideas seem intuitively appealing; the typical mode of discourse is argument" (Clinchy 1996, 207).

Bruner (1986) describes a similar way of thinking using the term "logico-scientific" or "paradigmatic" mode of thought. This mode "attempts to fulfill the ideal of a formal, mathematical system of description and explanation. . . . The imaginative application of the paradigmatic mode leads to good theory, tight analysis, logical proof, sound argument and empirical discovery guided by reasoned hypothesis" (12–13). Both separate and paradigmatic ways of knowing attempt to search for truths by excising the knower's experiences, intuitions, and emotions from the search and are often viewed as a means to achieve objective knowledge. Objective knowledge is thought to be untainted by personal experience and idiosyncrasies. Experiential modes of knowing are generally viewed as subjective. Subjective knowing is presumed to be less reliable, particularly in mathematics, which is often thought to provide access to objective knowledge.

Connected Knowing and Narrative Thought

> We learn mathematics by doing computations and other, less routine, constructions and by being surprised by the results of those constructions. Surely this is naturally described as learning from experience. (Goodman 1993, 64)

While separate knowing describes a thinking process whereby a person is said to extricate or separate his or her self from an idea and seek truth through a process of reason, an alternative process is to come to an understanding of a situation by appealing to personal experience. This form of knowing has been described by Belenky et al. (1986) as "connected knowing" and by Bruner (1986) as "narrative knowing." Connected and narrative knowing invite "images, models, metaphors, and even narratives" to form a holistic vision of the topic of concern (Elbow 1986, 264). Rather than determining truth by appealing to an external source or an independent reality, a connected knower asks the question, "What in your experience has led you to that point of view?" (Clinchy 1996, 206). A discourse drawing on "connected knowing" does not emphasize a process of doubting the other's idea or position or imposing a person's own ideas on another, but this discourse emphasizes that one tries to believe and understand the statements made by the other person, regardless of whether one agrees or disagrees with the statements made. Just as "doubting" can be described as a method for determining what is thought to be true even when the idea seems appealing, "'believing' is a procedure that guides her interaction with other minds; it is not the result of the interaction" (209).

Although I have described them independently, separate/connected, doubt/belief, and paradigmatic/narrative ways of knowing or coming to know are not dichotomies in which one is more correct or better than another. They coexist within persons as complementary pairs. Different situations or circumstances often call for different ways of coming to know. If we distinguish between knowing through doubting and knowing through believing as different processes for understanding, engaging in either a discourse of doubt or of belief is likely to lead to different forms of understanding. The understanding or knowledge that becomes valued is often dependent on the process of knowing that is expected in a particular situation. Mathematics generally emphasizes a methodology of doubt that is congruent with an argumentative discourse in mathematics education. In an inquiry-based classroom, students are often asked to scrutinize solutions offered by other students, and then, through a process of doubt or argumentation, agreement can be reached as to which solutions are more sophisticated or efficient, and what is or is not mathematically acceptable in terms of explanation and justification (Yackel and Cobb 1996). The subject matter of these exchanges is generally in relation to the practice of problem solving, its procedures, strategies, and solutions, and the explanations of these aspects of practice. The chapter that follows addresses the assumptions underlying what knowledge or ways of knowing our society values, and discusses the role of problem solving in present perspectives of mathematics learning.

Notes

1. This image was constructed by asking a number of educated adults to answer the following question: "What do mathematicians do—what image comes to mind?"

2. While the four-step approach and heuristics are general descriptors of what successful problem solvers might do, such heuristics have proved unsuccessful as prescriptive mechanisms for teaching problem solving (Putnam, Lampert and Peterson 1990).

Chapter 3

The Practice of Mathematics

What is Valued?

When we consider "What counts as knowledge?" we often provide answers which reveal underlying assumptions as to what knowledge or way of knowing is valued in the culture in which we are embedded. North American societies highly value scientific and mathematical thought. We believe that mathematics education, taught as a collection of specific concepts, procedures and processes, is essential knowing for students as future workers, if not for their particular job, then to provide access to well-paying careers, which are usually in the areas of science and technology (Noddings 1994). Valuing scientific and mathematical ways of knowing has a long history in western culture (Davis and Hersh 1986). Repercussions of this valuation are seen in the way mathematics is used as a filter for postsecondary entrance requirements, in public outcry over low scores on international achievement exams, and in the attention given to the lack of girls and women in science and mathematics. The same attention, fear, and sense of urgency are not associated with attempts to get more men into nursing or into producing more poets and musicians. Developing mathematical skills, particularly problem-solving skills, is viewed as essential for continuing our society's scientific and technological progress.

The Role of Problem Solving within the Practice of Mathematics

For the past two decades, problem solving has been the focus of the reform movement in mathematics education. In 1980, the NCTM's *Agenda for Action* stated that "problem solving should be the focus of school mathematics" (1). In 1989, this sentiment was reiterated in the NCTM *Standards* document. Recommendations for grades K-4 read:

> Problem-solving should be the central focus of the school curriculum. As such, it is a primary goal of all mathematics instruction and an integral part of all mathematical activity. (23)

Similarly, the revised *Principles and Standards* document (NCTM 2000) states that, "Solving problems is not only a goal of learning mathematics but also a major means of doing so" (52).

Lester (1994) writes, "it is safe to say that since the publication of the *Agenda*, problem solving has been the most written about, but possibly the least understood, topic in the mathematics curriculum" (661). Lester provides an overview of problem-solving research over a 25-year span which includes research emphases on, for example, "identification of characteristics of successful problem solvers" (1970–1982); "comparison of successful and unsuccessful problem solvers" (1978–1985); "relation of affects/beliefs to problem solving; metacognition training" (1982–1990); and "problem solving in context" (1990–1994) (664). He suggests "it is probably also safe to say that most mathematics educators agree that the development of students' problem-solving abilities is a primary focus of instruction" (661); however, he does not address why this belief is prevalent. Why *should* problem solving be the focus of instruction?

One response is implied in the grades 9–12 *Standards* document recommendations, which state that "mathematical problem-solving, in its broadest sense, is nearly synonymous with doing mathematics" (NCTM 1989, 137). The perception that problem solving and doing mathematics are synonymous has implications for how mathematical practice is represented and implemented in schools.

The assumption that mathematics *is* problem solving is one possible reason why problem solving is viewed as a necessary focus in school curricula. Two other possible assumptions, discussed below, are that problem solving is perceived as an important skill for persons living in a technological world, and that successful problem solving is also synonymous with intelligent behavior.

Problem Solving and Technology
The importance of problem-solving abilities for children is intimately tied to the perception of and desire for an increasingly technological society. Curricular documents commonly promote beliefs surrounding the utility of mathematics for solving problems in today's technological world and the importance of developing proficient problem solvers for future purposes and job prospects.

While the educational goals are not restricted to developing problem-solving skills for utilitarian purposes related to society's technological needs, they receive priority and reflect the perceived needs of society:

> The need to understand and be able to use mathematics in everyday life and in the workplace has never been greater and will continue to increase. . . . The underpinnings of everyday life are increasingly mathematical and technological. (NCTM 2000, 4)

Problem solving as an instructional focus seems to provide a means for addressing this concern.

However, mathematics cannot always be linked to such practical purposes. Educators, claims Tymoczko (1993), often make the mistake of stressing the utility of mathematics even when its practical purpose is negligible. Engaging in mathematics as a practice is often driven more by passion and curiosity than utility. Mathematics is an emotional/intellectual enterprise and the teaching of mathematics must be cautious not to ignore the psychological components of doing mathematics. By placing an emphasis on utility, Tymoczko says educators

> often hide from their students the excitement and intrinsic interest of mathematics: they hide behind a facade of supposed utility. It's rather like trying to awaken people to the joy of mountain climbing by trying to convince them that someday they might need to climb a mountain. (11)

Emphasizing the usefulness of mathematics for a technological society is one assumption that has increased the importance of problem solving in school mathematics. Although this goal is not detrimental to learning mathematics in and of itself, if problem solving is viewed as only or as primarily a utilitarian venture, the field of acceptable mathematical activity or mathematical practice in schools is narrowed. The question, "When will I ever have to use this?" will continue to be asked, and teachers will continue to try to search for practical examples to (inadequately) affirm the usefulness of a mathematical topic or algorithm. This description of problem solving activity which bounds our understanding of knowledge acquisition and acceptable mathematical practices in classrooms has further implications for describing *intelligent* problem-solving behavior.

The next assumption that is addressed for examining the place of problem solving in the curriculum is the tendency to equate problem solving with intelligent action.

Problem Solving as Intelligent Behavior

A commonly cited definition for a mathematical problem is provided by Charles and Lester (1982): A problem is a task for which the person confronted with it (1) wants or needs to find a solution, (2) does not have a readily available procedure for solving it, and (3) needs to expend effort to find a solution. This definition remains relatively constant under different descriptions of problem-solving activity, but the process by which one solves the problem is described differently depending on the theory of cognition under which the behavior is being observed. Two learning theories currently in use in our mathematics classrooms are cognitivism and constructivism.[1] Problem solving plays a prominent role in both of these perspectives.

From a cognitivist perspective, problem solving begins with a well-defined "task environment" in which all possible states or actions within the environment can be predetermined (Winograd and Flores 1986). The process a problem solver follows is first to internally represent the environment through the use of symbol structures, then, process the symbolic information to analyze courses of action within the environment, and finally, select the course of action which best achieves the desired goal. In this view, cognition is modelled on an input-output or computer metaphor. Intelligent behavior is said to occur "when symbols appropriately represent some aspect of the real world, and the information processing leads to a successful solution of the problem given the system" (Varela, Thompson, and Rosch 1991, 42–43). Thus, any input such as a question, task or activity which produces predictable outputs or answers is potentially defined as a problem. From this perspective, mathematically intelligent behavior is said to occur if a student produces an answer to a problem which accurately matches or represents the environment. Problem solving as described under cognitivism and the related perception of intelligent behavior provides a framework for mathematical practice in schools which is prevalent in many classrooms today. Problems within a cognitivist perspective are used as inputs. They are expected to be extremely well defined so that outputs can be predicted ahead of time. Incorrect outputs can be "diagnosed" so that appropriate instruction can follow. This perspective is also supported by a common discourse pattern described by Sinclair and Coulthard (1975) as "Initiation-Response-Feedback." This I-R-F pattern of discourse is played out every time a teacher poses a question usually of a well-defined nature (input); a student provides a response generally in the form of a numerical answer (output);

and the teacher provides immediate feedback as to whether that response is correct or incorrect according to a predetermined solution. The discourse and mathematical practice described under cognitivism is interrelated with how a person views or experiences truth and reality. Truth, and in this case mathematical knowledge, is viewed as an accurate representation of a pregiven reality.

Problem solving under a constructivist perspective provides a different description of intelligent behavior, and subsequently a different perspective on truth and reality. "Knowledge" as accurate representations of reality, is replaced with "knowing" as a process of reconstructing and reorganizing conceptual structures (von Glasersfeld 1995b). Similarly, *truth* is replaced with *viability* or *fit* in reference to one's physical actions and experiences. That is, "what determines the value of the conceptual structures is their experiential adequacy, their goodness of *fit* with experience, their *viability* as a means for the solving of problems" (von Glasersfeld 1987, 5). Intelligent action is not prescriptive as was described in cognitivism; instead constructivism suggests that intelligent behavior follows a *proscriptive* logic: any action which is feasible or possible within a given setting is acceptable. The locus of knowledge or knowing and the authority for determining whether the knowledge is acceptable shifts from "out there," from matching representations of an external reality, to "in here," subjectively determined in the mind of an individual. Constructivism contends that it is the person, in interaction with other persons and the environment, who determines whether actions are viable or intelligent.

The teacher's role under a constructivist perspective of knowing is to design instructional activities meant to be problematic for students. Such activities provide opportunities for students to reorganize their action and potentially reorganize conceptual structures to resolve the problems encountered (Cobb, Yackel, and Wood 1995). Problems in this setting are viewed as potential sources of cognitive conflict or perturbation. Learning under a constructivist perspective is characterized as a process of eliminating perturbations through conceptual reorganization (Cobb 1994). "To solve a problem intelligently. . .one must see it as an obstacle that obstructs one's progress towards a goal" (von Glasersfeld 1995a, 14). From a constructivist view, problems are viewed as containing obstacles to overcome.

An I-R-F discourse is inappropriate under constructivism. In its place, discourse which allows for whole class and small group discussions has the potential of exposing differing conceptual structures between students.

Discussions potentially give rise to perturbations or cognitive conflicts through which learning may occur. Providing opportunities for argument encourages students not simply to accept the ideas presented, but to negotiate through processes of argumentation, justification and counterexamples which solution is most viable for themselves.

From a teacher's and an observational perspective, conflict as it might occur in argument may be viewed as a mechanism for instigating and observing learning. As conflicts are an internal phenomenon, the mechanism used to identify instances of perturbations are the arguments that arise between students resulting from differing ideas and understandings. Studies such as those conducted by Yackel, Cobb and Wood (1991; Cobb, Yackel and Wood 1992; Yackel and Cobb 1996) have attempted to identify sources of learning by providing opportunities for cognitive conflicts in small group interactions.

From both a cognitivist and a constructivist perspective, problems are seen as an important source for observing and provoking mathematical learning either through input/output processes or as sources of perturbations. The previous discussion on constructivism presents a view of problem solving as a practice of engaging in activities, questions, tasks, or problems in an effort to eliminate perturbations. A discourse of argumentation in conjunction with a practice of problem solving provides additional sources of perturbations through social interaction.

While problem solving as a mathematical practice within a discourse of disagreement has the potential to be a rich source of mathematical activity and learning, I question whether this image of mathematical practice in schools is broad enough and, if it is not, whether argumentation is always the most appropriate form of discourse for engaging in mathematical activity with others. While problems have expanded from their previous implementation as word equations, teachers and students interacting within either a framework of cognitivism or constructivism, often continue to view problems as discrete and separate tasks presently redesigned to stress the usefulness of mathematics in today's world. Although the number of tasks has been drastically reduced, students are still expected to complete a series of problems one after another. The goal for problem solving, as it was for drill and practice exercises, remains on completing tasks as quickly and as efficiently as possible so that the students can be "released" from their relationship with the mathematics. I question whether it is wise or sufficient to equate mathematics or mathematical practice solely with problem solving. I also wonder whether there are other forms of legitimate practice that may encourage students to maintain their relationship with the mathematics. However, for an alternative practice to be accepted it must be viewed as appropriate and intelligent by persons presently in the practice of mathematics. The following section uses recent literature in the practice of mathematics to broaden our image of legitimate mathematical activities beyond problem solving and beyond proof.

Experiment/Experience in Mathematical Practice

> The definition-theorem-proof approach to mathematics has become almost the sole paradigm of mathematical exposition and advanced instruction. Of course, this is not the way mathematics is created, propagated, or even understood. The logical analysis of mathematics, which reduces a proof to an (in principle) mechanizable procedure, is a hypothetical possibility, which is never realized in full. Mathematics is a human activity, and the formal-logical account of mathematics is only a fiction; mathematics itself is to be found in the active practice of mathematicians. (Davis and Hersh 1981, 136)

Ernest (1994) states that mathematics is conversational at its base. In order to make this claim however, a number of fundamental principles need to be rejected:

1. there is a secure and fixed basis of truth on which mathematical knowledge is founded;
2. there are wholly reliable logical deductions of mathematical theorems from explicit premises;
3. absolute mathematical knowledge based on impeccable proofs is an ideal which is attainable;
4. the logical properties of mathematical proof alone suffice to establish mathematical knowledge without reference to human agency or the social domain. (35)

In order to reject these principles, a broader description of mathematical practice is required. Recent developments in the philosophy of mathematics support the rejection of these principles by broadening what counts as legitimate mathematical activity—activity which includes a mathematician's initial states of exploration, rather than simply focusing on the mathematical abstractions and proofs resulting from these explorations and tentative beginnings. Polya (1954) described this aspect of practice under a process he termed "plausible reasoning" discussed in the previous chapter. While Polya and Lakatos started us in a direction that allows us to challenge the principles stated above by questioning the nature of mathematics and its practices, their descriptions of mathematics have not taken into account the development of mathematics over the past fifty years. New areas in mathematics which emphasize experimental approaches have surfaced. However, even these experimental mathematicians are faced with criticism as their 'brand' of mathematics is often viewed as lacking rigor and, hence, is not considered real mathematics. Yet, the definition of mathematics and what counts as legitimate mathematical activity is not, or should not be historically static. As Davis and Hersh (1981) write, "the definition of mathematics changes. Each generation and each thoughtful mathematician within a generation formulates a definition according to his lights" (8). Until a broader definition of mathematics is accepted, it will be difficult to value the roles that experiential and narrative ways of knowing have in mathematical practice. Although less valued, they are an essential part of such practice:

> Many scientific and mathematical hypotheses start their lives as little stories or metaphors, but they reach their scientific maturity by a process of conversion into verifiability, formal or empirical, and their power at maturity does not rest upon their dramatic origins. Hypothesis creation (in contrast to hypothesis testing) remains a tantalizing mystery—so much so that sober philosophers of science, like Karl Popper, characterize science as consisting principally of the falsification of hypotheses, no matter the source whence the hypothesis has come. (Bruner 1986, 12)

In hypothesis creation, the narrative artifacts arising from personal experience, including "little stories" and "metaphors," are used as "crutches to help us get up the abstract mountain. Once up, we throw them away (even hide them) in favor of a formal, logically consistent theory that (with luck) can be stated in mathematical or near-mathematical terms" (Bruner 1986, 48). While it is presently more common to acknowledge the importance of experience in generating understanding and admit to the fallibility of mathematical knowledge *in principle*, there is still a tendency in the practice of mathematics to devalue and hide the stories and metaphors behind mathematical hypotheses and erase errors along the way so that mathematics may be presented in an "eternal" form. On the subject of errors, Hadamard (1945/1973) wrote that

> Good mathematicians, when they make [errors], which is not infrequent, soon perceive and correct them. As for me (and mine is the case of many mathematicians), I make many more of them than my students do; only I always correct them so that no trace of them remains in the final result. (49)

Ignoring, denying or suppressing the "crutches" of mathematical understanding and the errors of activity is not surprising given our long history of believing that mathematics provides access to universal truth. The only way in which to achieve truth which stands apart from human existence is to deny or ignore the human processes involved.

> Mathematics, one of the most imaginative activities in the history of humankind, has been supplied with foundations that seek not to explicate mathematical imagination precisely, but rather to ban from the characterization of mathematics all forms of imagination—images, metaphorical thought, signs, pictures, narrative forms. Of course, practicing mathematicians go on using their minds with all of their unconscious imaginative mechanisms working silently, unobserved, and undescribed in the highest of high gears. (Lakoff and Núñez 1997, 29)

Many of the human aspects of mathematical practice have been ignored in discussions of mathematicians and by mathematicians of their work. Recently, philosophers of mathematics have criticized the immaculate form in which mathematics is generally presented and have drawn on recent developments in mathematics to reveal the messiness and experimental nature of some mathematical practices. Epstein and Levy (1995), in promoting their journal *Experimental Mathematics*, state that the format for a published mathematics paper is disturbing:

> What one generally gets in print is a daunting logical cliff that only an experienced mountaineer might attempt to scale, and even then only with special equipment. Is this the best thing for the research community? Is it fair to graduate students? Should we give the impression that the best mathematics is some sort of magic conjured out of thin air by extraordinary people when it is actually the result of hard work and of intuition built on the study of many special cases? (670)

The lack of acknowledgement of narrative aspects and experiential ways of engaging in mathematical activity has upheld the mystery surrounding mathematics inquiry. The frame placed around acceptable mathematical practices gives the appearance that only the intellectually elite are capable of achieving understanding in mathematical domains of knowing. The frame keeps the processes and practices of mathematical inquiry hidden from its students of all ages.

Although Lakatos emphasized the role of experimentation and observation of examples *in* mathematical activity, mathematics *as* observation and experimentation are often contested as acceptable forms of mathematical practice. Epstein and Levy (1995) defend experimentation and the importance of remaining grounded in data and examples: "Most mathematicians spend a lot of time thinking about and analyzing particular examples. . . . Gauss declared, and his notebooks attest to it, that his way of arriving at mathematical truths was 'through systematic experimentation'" (670).

It is interesting to note that experience and experiment have the same etymological root, *experiri*, which means "to try or test" (O.E.D. 1992). One of the two meanings for 'prove' is also related to trying and testing. The familiar definition, "establish beyond doubt," which implies a process of methodological doubt referred to earlier, is generally viewed as the primary activity of mathematicians, but proof also stems from the meaning "to try or test." Although archaic, it is still found today in such terms as a "printer's proof" (Epstein and Levy 1995). Perhaps to broaden our image of mathematical practice this definition needs to be resurrected. Trial-and-error methods of testing in mathematical experiments are frequently conducted on computer while other methods are still the result of pencil-and-paper work or building physical models (Epstein, Levy, and de la Llave 1996). Regardless of the tool used, experiments are tentative procedures in which knowledge, or rather knowing, is gained through processes of trying and testing through active experiment. Experimental and experiential aspects have always been prevalent in mathematics; however, these aspects, when they are explicitly used to explain mathematical phenomena push the boundaries of what has traditionally been accepted as legitimate mathematical practices. For example, computer experiments are commonly used within the study of chaotic systems. The research methods used made experimentation much more visible within the practice of mathematics: "Those who used computers to conduct experiments became more like laboratory scientists, playing by rules that allowed discovery without the usual theorem-proof, theorem-proof of the standard mathematics paper" (112). Other fields of mathematics that have originated and grown through the use of experimental approaches include investigations of minimal surfaces such as those found on soap bubbles and films, growth of crystals, and fractals. In these areas of research, "investigators test their ideas by representing them graphically and doing calculations on computers" (Horgan 1993, 95). Investigations involving computer experiments and proofs do not adhere to the standards for rigorous proof in mathematics because they are not, for instance, "surveyable" by humans. As a result their reliability and objectivity are often called into question (Kleiner and Movshovitz-Hadar 1997).

The criticism against experimental mathematics is based on a perspective of how mathematical problems should be proved. Although Horgan (1993) suggested that the development of experimental approaches called for "The Death of Proof," his point was to emphasize that both human-made *and* computer-made proofs are tools of mathematicians to *explain* the phenomena they are investigating. From this perspective, proof is viewed as a form of explanation or as Hersh (1993) suggests, "*a proof is just a convincing argument, as judged by competent judges*" (389).

Rejection by many people of experimental mathematics *as* mathematics is not necessarily because experimental proofs do not provide convincing arguments, but because the arguments have not been formulated according to the criteria set by the judges. For these people, *mathematics is proof* in which the description of proof defines the practice of mathematics and the discourse in which the results are presented. Similarly, as stated in the previous chapter, many schools are currently promoting the notion that *mathematics is problem solving*. The descriptions provided in the NCTM and curriculum documents also describe a form of mathematical practice and related discourse. Both of these statements regarding the nature of mathematics contain underlying assumptions about what mathematics is valued, what are acceptable mathematical practices, and how the results of those practices should be shared with others in the community. Experimental mathematics is controversial under the statement "mathematics is proof," and the statement, "mathematics is problem solving" is perhaps not broad enough to encompass experimental mathematics within its description of mathematical practice. In the practice of proof and problem solving, the nature of mathematics arises in how one views and interacts with the body of mathematics.

The *Body* of Mathematics

Mathematical investigations, experimental or otherwise, contain their own sets of interesting problems and proofs of solutions to these problems. But these problems are generally not preformulated and offered to the mathematician to solve or prove. Rather, problems arise as questions or concerns through the mathematician's ongoing curiosity as he or she interacts with the phenomenon. Conjectures as "tentative" explanations may be formulated in response to one's interactive experiences. The explanations offered may give rise to new questions and conjectures or may alter previously posed questions, conjectures, and explanations. The new questions and conjectures at any moment in time map out an area of study as an interconnected network of problems. Yet, these problems are not static, nor are they isolated. A problem is continually reformulated each time it is addressed as information from connected problems alters how the mathematician now might pose the question from which the problem originally arose and how he or she might proceed in solving it.

Or a solution to one problem may become a source for new questions which may expand the network further. Therefore, solving one problem has implications for how other problems are perceived and solved within the network, or at least the problems connected to it in some way. The problems themselves change and evolve. The whole network constituted by these problems changes and evolves. The network becomes a pulsating "body" where activity in one area has implications for and effects on other activity within the network. At any time, however, we can carve out one problem in the network, focus in on a mathematician's activities in direct relationship to that problem, and allow the network to fade into the background. By doing so, we selectively ignore how the problem arose as a problem—that is, we ignore the history of the problem embedded in the network, and we do not attend to the experiences the mathematician brought to the problem in his or her engagement in other parts of the network. Under this focused view of mathematical activity, experimental mathematics would likely fit under the description of "mathematics is problem solving." Under a broader image of the "body" of mathematics as an interconnected, evolving and expanding network, problem solving appears to be limiting.

Every field of study in mathematics may be viewed as an evolving and networked body of questions, problems, conjectures, and explanations. Activity in one area of a mathematical network arises from activity in other areas, and may affect subsequent activities in connected parts of the network. Yet, these fields of study or networks are themselves interconnected, forming larger networks which are interconnected again and again within larger networks. At some point, at some level, these interconnected networks form the *body of mathematics*. The fractal nature of the body of mathematics, made up of evolving networks on many levels, presents an image of mathematics as a dynamic living system. From each level, we can see the body of mathematics on different scales—as an interconnected set of problems or human concerns. On a larger scale we can imagine that geometry, calculus, and algebra form major points of intersection within the networked body of mathematics. We can then select and bring into focus any one of these points, such as geometry. Within geometry another network arises, including points for Euclidean geometry, fractal geometry, and topology. We can refocus our observational lens again and again to smaller and smaller self-similar images such that one proof for one problem is itself a network of interconnected problems. It is at this level that Lakatos (1976) provided us with an image of the interconnected network of Euler's formula. While we could look at the proof of Euler's theory as a final product, Lakatos demonstrated how the history of mathematics is gathered into the moment of the final proof. While a mathematician may spend an entire lifetime working on only one of its problems, he or she is likely drawing upon and potentially affecting a large magnitude of networks within the body of mathematics.

It is not always useful to imagine the whole of the evolving body of mathematics at once. At times, it may be important to allow much of it to fade into the background so that we may focus on one point, or one specific problem within a network. An argumentative discourse is useful for carving out a problem and focusing on and formulating convincing arguments or proofs for the individual problems within a particular network or 'body part' within the body of mathematics. Argumentation allows a community to develop and select explanations which best help individuals understand a present and particular concern. In relation to proof, Hersh (1993) states that "proof is a *complete explanation*. It should be given when complete explanation is more appropriate than incomplete explanation or no explanation" (397, italics added). While a proof which meets the criteria defined by the discipline of mathematics is not always an appropriate goal within classroom instruction, a discourse of argumentation in classrooms focuses on the necessary practice of providing a "convincing argument" and a "complete explanation" for individual problems.

When we expand our focus and allow the ever-evolving body of mathematics into our sights, it becomes difficult to imagine that a complete explanation is possible. When one brings forth this image of mathematics, all explanations are necessarily incomplete. Perhaps a different form of discourse which attends to and allows for interconnections between problems and between networks is necessary when the body of mathematics is viewed as a dynamic living system. While argumentation is a suitable discourse for removing the "crutches," errors, and imagery of mathematical explanations, it is likely to be less suitable for creating the narratives and metaphors that initially aid understanding, for expanding the boundaries of a mathematical network to new questions and concerns, and for reformulating and revisiting previous conjectures in response to the expansion. A different form of discourse is likely necessary to support these aspects of mathematical practice.

A Place for Conversation within the Practice of Mathematics

The aim of this work is primarily to explore the place for conversation within the practice of mathematics. Prior to a discussion of conversation as a form of interaction, one must be willing to expand one's view of mathematical practice to encompass activities beyond problem solving and beyond proof. Mathematical conversation requires alternate images of practice, intelligent action, acceptable explanation, and the nature of mathematics.

However, this is not an easy task. The body of mathematics is often viewed as an existing set of truths, separate from human interaction and referenced on an external *universe* as was described in cognitivism. This image of mathematics arises even when our intentions are to view mathematical knowing as a personally constructed activity. From this perspective, the body of mathematics is an external entity whereby mathematical explanations are judged by how well they appear to match reality and achieve universality.

While a cognitivist perspective of reality has been called into question, the alternative offered by constructivism is a cause for concern for many teachers. Constructivism offers a *multiversal* view of reality. Validation of explanations is based not on consistent and certain truths but by referencing one's own experiences (Maturana 1988). If individuals construct their own understanding and determine their own viability of that understanding, then many mathematical explanations become potentially legitimate. This image creates an emotional discomfort for many people, as it challenges their perceptions about what mathematics is and how it is validated.

Neither a cognitivist nor a constructivist view of truth and reality are "correct." They are simply explanations for experiences we have as humans. At times it may be more appropriate to choose or act within one set of explanations or another, depending on one's emotions, how one perceives the situation at hand and how one observes activities within the situation. A *conversarial* reality, offered below, is another explanatory alternative. It attempts to find a way in the middle—to relieve some of the tensions experienced when offering or viewing explanations within a *uni*verse with one certainty and a *multi*verse with uncertainty (Steier 1995).

A Conversarial Reality

A conversarial reality is drawn primarily from a constructivist or a multiversal view. Within this view, however, an emphasis is placed on the notion that our experiences do not occur in isolation. In observing mathematical activity as part of our ongoing acts of living, our gaze expands to include the ongoing interactions, the ongoing conversations with others and the mathematical environment. Such conversations are also embedded in a long and rich history and culture. Our experiences and our explanations of our experiences occur and are expressed in conversation.

Given that we are similar biological and social beings, our experiences and our ways of perceiving our experiences will be similar. A person's explanations, then, are not random nor are they viewed as subjective within a conversarial perspective. A limitless number of explanations is not imaginable for any human experience. They are bounded by the physical and perceptual or phenomenological experiences we share as humans. Therefore, we bring forth common and overlapping cognitive realities. A multiversal reality seems reasonable when individuals are viewed to exist autonomously in isolation from one another. But since humans exist in conversation, a conversarial reality is an alternative that emphasizes the importance of our interactions and the ethical and moral responsibilities that we have to one another through these interactions.

Explanations within a conversarial reality are not brought forth through a process of intersubjectivity, for intersubjectivity suggests that there is a subjective/objective dichotomy and that resorting to our subjective or intersubjective explanations is necessary because of limitations in our cognitive capacities as humans to know the real world out there. Instead, this perspective proposes that our ongoing effort to explain is simply a condition of our existence; from our historical origins to the present, as humans we engage in an ongoing and evolving process of finding coherence and meaning in our lives (Maturana 1998). Within a conversarial reality then, explanations, mathematical or otherwise, cannot be verified, falsified or confirmed. Explanations are simply offerings to ourselves and others to provide meaning for our experiences at a particular moment in time.

As we converse and share our explanations with others through recurrent conversations, our explanations of our experiences tend to "coemerge" (Kieren, Gordon Calvert, Reid and Simmt 1995). We perceive reality similarly not because it is external, nor because we are enculturated into a particular reality, but as we coexist in conversation with others, *objects* are commonly/communally brought forth and distinguished; therefore, we bring forth a reality together through a process of "inter-objectivity" (Maturana 1998). The distinctions of objects, ideas, problems, and so on that we create in coexistence with others bring forth common or overlapping realities that are lived on a moment-to-moment basis.

Reality of Choice

The reality one lives or perceives is not fixed. It is brought forth through the ways in which one formulates his or her explanations and offers them to others. At any moment one may bring forth a universal, a multiversal, or a conversarial reality. The reality lived changes as one's activities change and as the people one interacts with change. However, the reality brought forth is not simply chosen or determined through a process of reason; it arises from an emotional predisposition occurring in the moment of interaction (Maturana 1988). One's emotions at a particular time determine how one frames and validates or justifies his or her explanations. It is in the validation of one's explanations that a reality is brought forth. Validation may occur by referencing an external reality to bring forth a universe, or by referencing actions and experiences to bring forth either a multiversal or conversarial reality.

If one brings forth a universe perceived as an external reality, one "sees reality as that which is" (Maturana 1988, 39). In mathematics we tend to bring forth a universe based on an ancestral history of recurrent human conversations that conserves the belief that mathematics provides access to universal truths. While mathematics has long been thought to be divorced from the realm of emotion, we can imagine the emotional power that mathematics provides to a society who feels that mathematics is capable of touching "the sphere of the Divine" (Triadafillidis 1998, 22). Explanations formulated and offered within this emotional grounding bring forth a universal reality in which certainty and truth exist and are obtainable through reason. Different truths for the same event or idea are not possible, so when alternative explanations arise, interaction proceeds under an emotional desire to compel others to see truth by attempting to *convince* them of their erroneous visions.[2] While the bringing forth of a universe cannot be viewed as a correct or incorrect depiction of the way things are, there are emotional consequences for interaction within this view.

From this perspective, a person may come to believe that he or she can know truth through a process of reasoning in which the powers of reasoning are given cognitive properties contained within the individual. If a person is convinced that he or she has achieved truth, the other person in interaction serves no purpose other than to be convinced of this truth as well. In effect, the other, along with his or her explanations and experiences, may be negated. Only one explanation can be correct or more accurate; therefore, attempting to understand the other's explanation is not necessary for the person who feels he or she has achieved truth. Within a universal reality there is a fear that anytime a person offers an explanation that he or she is uncertain of, it will be rejected by the community. When it is rejected, the experiences leading to that explanation become irrelevant and are ignored. An incorrect explanation is inadvertently a comment on a person's given cognitive abilities to gain access to truth; that is, his or her intelligence may be questioned. Therefore, a rejection of an explanation is often felt as the negation of one's personal self and his or her experiences and intellectual capacities, not just of one's ideas. In these circumstances, rejection of an explanation is often felt as an "existential threat" (Maturana 1988, 63).

If we recall Lampert's students' responses to participating in a discourse of disagreement, their reactions were emotional, and rejection of explanations were perceived as threats to their personhoods and existence. One student's comment was, "it can get sort of embarrassing at times . . . And it makes you sort of feel like you want to crawl into a hole and die" (Lampert, Rittenhouse, and Crumbaugh 1996, 742). Although discouraging, the students' emotional reactions are consistent with bringing forth a universe in which only one correct answer is possible and a rejection of a student's answer may be felt as a rejection of the person him- or herself.

If one brings forth a multiverse, explanations are validated through experiences rather than by referencing an external reality. When a multiverse is brought forth, Maturana states that "all views, all verses in the multiverse are equally valid. Understanding this, you lose the passion for changing the other" (Maturana 1985). The emotional predisposition alters from the desire to convince one another to an acceptance of the other and his or her experiences and explanations. However, a multiverse is less satisfying in mathematics. We have a great deal of difficulty accepting that all explanations that maintain coherence within the domain of mathematics might be equally valid. A conversarial reality is an alternative that is brought forth when one acknowledges the interactional aspects of human existence. *Understanding* rather than convincing the other person becomes the emotional grounding for stating explanations within a conversation. The desire to broaden one's understanding of a mathematical phenomenon is aided by listening to and trying to understand any and all experiences and explanations of the other. This emotional grounding for bringing forth a conversarial reality constitutes a relationship also based on mutual acceptance rather than negation. Differing explanations and disagreements between persons provide opportunities for both persons in conversation to broaden their understanding. Since there is no search for a single ultimate explanation, the desire to change the other's viewpoint is not felt or demanded. The purpose in conversation is not to impose our views on the other or to find the "correct" answer, but to open ourselves to the possibilities of new understanding for the experiences we share. In the bringing forth of a conversarial reality, our explanations, including our mathematical explanations, tend to coemerge through our recurrent interactions.

The focus of this work is to broaden our understanding of what can be counted as legitimate mathematical practice and mathematical knowing. It recognizes mathematics as a human activity that is in part a practical problem-solving pursuit, but also speaking more broadly, a historical and evolving bodily network brought forth through people in conversation. Mathematical conversations attempt to account for the emotional, embodied, interactive, and ongoing experimental endeavors of mathematical activity. Broadening what we consider to be legitimate mathematical practice, however, requires effort to adjust the emotional basis for validating explanations from attempts to convince the other, to accepting and trying to understand the other's experiences and explanations. Chapter 4 provides an in-depth look at conversation as it is presently used in mathematics education, but then moves beyond this local use to a broader theoretical and philosophical basis.

Notes

1. There are a wide variety of constructivisms. The term is used here to refer to "radical constructivism" as described by von Glasersfeld.

2. The etymology for the word *convince* is Latin stemming from the prefix "con" meaning *altogether* or *wholly* and "vincere" meaning to *conquer* or *overcome* (O.E.D. 1992). Such a definition gives further credence to the metaphor "argument is war" and the belief that arguments are won and lost.

Chapter 4

Creating Space for Mathematical Conversations

Moving Beyond an Impoverished View of Mathematical Conversation

The pursuit of learning is not a race in which the competitors jockey for the best place, it is not even an argument or a symposium; it is a conversation. . . . A conversation does not need a chairman, it has no predetermined course, we do not ask what it is 'for' and we do not judge its excellence by its conclusion; it has no conclusion, but is always put by for another day. (Oakeshott, 1989, 8)

Mathematical conversation evokes an image of open communication between teacher and students and among the students themselves. It challenges the image of a silent mathematics classroom where the talking by students has been limited to one word or numerical responses to teacher-directed questions following an Initiation-Response-Feedback pattern of discourse (Sinclair and Coulthard 1975). This chapter examines the present focus on conversation as a means for providing opportunities for students to learn to "talk" mathematics. It is currently recommended in mathematics educational reforms that "students have the opportunity to read and discuss ideas in which their use of the language of mathematics becomes natural" (NCTM 1986, 6). Following this discussion is an attempt to expand on the notion of conversation as talk to provide a broader basis of mathematical discourse and practice.

The term "mathematical conversations" occurs frequently in mathematics education literature. Its use is perhaps a way to address the current emphasis on communication in the classroom. In one chapter titled, "Fostering Mathematical Conversations," Whitin and Wilde (1995) use the term "conversation" to describe whole-class discussions around a children's literature book relating to mathematical ideas. Similarly, Goldenberg (1991) writes an article titled, "A Mathematical Conversation with Fourth Graders," in which transcripts of teacher-student whole-class discussions about decimal numbers are described. An approach similar to Lampert's discourse of disagreement is described in an article titled, "Listening to Students: The Power of Mathematical Conversations" (Atkins 1999). Whole-class discussions are initiated after students are asked to give a "thumbs up" or "thumbs down" to an answer provided by a fellow student.

From a research perspective, conversation has been emphasized for its potential to enhance learning. In Haroutunian-Gordon and Tartakoff's (1996) article "On the Learning of Mathematics through Conversation," they describe conversation as a form of "interpretive discussion." Conversation here is also described as whole-class discussions, but the authors state that what is said or the problems posed arise from these conversations: "The discussion of mathematics flows out of the ideas, beliefs, and concerns expressed by the students, and the topics pursued are the ones they needed to pursue in order to further their understanding" (4). Through their teaching experiment they pose that "mathematics, like so many other parts of life and domains of understanding, can be explored in a natural way" (8).

Sfard, Nesher, Streefland, Cobb and Mason (1998) discuss the question: "Learning Mathematics Through Conversation: Is It as Good as They Say?" Each author presents a slightly different view of conversation; however, it is generally used as a synonym for talking mathematics and communicating mathematically. Within the article, Nesher describes conversations among mathematicians as engaging in a discourse of argumentation within professional journals and at conferences. She then suggests that children could engage in parallel conversations in which they learn what is a *convincing argument* in mathematics. Nesher suggests that this could become the core of our mathematics teaching. The description and use of conversation here is very similar to the discourse of argumentation described in chapter 2.

In almost every instance, conversation is embedded in a constructivist framework and is described as a type of whole-class discussion in which the students are expected to explain their thinking. The teacher is seen as the facilitator of discussions, posing questions to encourage students to clarify their thoughts for the class and to expose conflicting ideas. Although this form of whole-class interaction is important within mathematics education, the image of *conversation as a whole-class discussion* did not fit with my original conception of "genuine conversations" in which conversation was a form of interpersonal interaction. The meaningful and thoughtful conversations I have with colleagues and friends are generally not facilitated discussions or arguments, but rather are intimate interactions over some topic of mutual concern. I see conversation as

> voices co-versing in prose, in poetry.
> Bodies rhythming in language, in gesture.

Conversation carries with it a sense of embodiment, presence, responsiveness, and responsibility. In our conversations with others we explore a topic of mutual interest and at the same time further our relationship with each other.

It has only been in recent history that conversation has been defined as a form of talk. The etymology of conversation (O.E.D. 1992) opens this word to richer, communal meanings:

Conversari (Latin)
   "to dwell in or with" others
   entering into "a spiritual or mental communion" with others.
Converse (Latin)
   "to turn"
   Two minds, two bodies, turning around each other; two perspectives entering into language turning over the words, returning to thoughts, ideas stories; transforming experience in mutual reciprocity.

Conversation carries with it "an implication of authenticity, of lived experience, of ideas coming alive in the context of everyday existence" (Simpson 1995, 49). However, conversations are not random or unpurposeful. Rather than predefined goals or end-states, conversations proceed on relevant concerns and directions that emerge in the moment and arise from the phenomenological history that each person brings to the conversation.

Investigating mathematical conversations from an interactive and embodied perspective may appear to many people as contradictory. Doing mathematics is often perceived as independent, linear, and goal-directed behavior, while a conversation rarely proceeds in a linear progressive manner with clear direction and purpose. Its character is more circular, weaving its way around and through the topic at hand (Smith 1991). The participants themselves do not know where they are going or even what they are talking about in some absolute sense. They are patient—waiting for future comments to make sense of incomplete or bewildering statements previously made. Meanings are constantly being shaped and reshaped while the topic is molded and transformed within the course of the conversation producing it. Gadamer (1989) puts it eloquently when he writes:

> We say that we "conduct" a conversation, but the more genuine a conversation is, the less its conduct lies within the will of either partner. Thus a genuine conversation is never the one that we wanted to conduct. Rather, it is generally more correct to say that we fall into conversation, or even that we become involved in it. The way one word follows another, with the conversation taking its own twists and reaching its own conclusion, may well be conducted in some way, but the partners conversing are far less the leaders of it than the led. No one knows in advance what will "come out" of a conversation. (383)

Philosophers and theorists in many fields (e.g., Bakhtin, Gadamer, Habermas, Maturana, Oakeshott, Rorty, Shotter, Vygotsky, and Wittgenstein) have characterized the significance of conversation as an ecological metaphor for human knowing, interaction and embeddedness in a physical, social, cultural and historical world. Rather than simply a "facilitated discussion," the work presented in this chapter attempts to provide a richer image and foundation for conversation by creating a theoretical framework which draws primarily from the work of Maturana, Gadamer and Bakhtin. The framework, labelled *enactivism*, attempts to relate conversation to interaction, understanding, language and reality.

Enactivism

> If there is no other, there will be no I. If there is no I, there will be none to make distinctions. (Chuang-tsu, 4th Cent., B.C. as cited in von Glasersfeld 1990)

Knowing through Interaction

Enactivism is a theory of knowing based primarily on the work of biologists Maturana and Varela (1987; Varela, Thompson and Rosch 1991). It interrelates contemporary cognitive science with philosophy, psychology, and ecological perspectives. The enactivist framework presented here also draws on Gadamer's "philosophical hermeneutics" (1989) and Bakhtin's "philosophical anthropology" (1981, 1986). Enactivism attempts to transcend many of the dichotomies that have existed in previous theories of mind, such as splits between subjectivity/objectivity, knowledge/action, human/world, and mind/body. What follows is a discussion of some of these dichotomies and enactivism's attempt to find a middle ground.

Cognitive theories generally attempt to answer the question "Where is the Mind?" (Cobb 1994). Cognitivists or representationalist models pose that knowledge is "out there" in the environment awaiting discovery. A radical constructivist, on the other hand, locates the mind "in here"—inside the head of an individual. This perspective claims that all that is knowable to an individual is constructed by him or her through subjective activity (von Glasersfeld 1988). While a cognitivist perspective privileges the environment, a constructivist perspective privileges the individual. Enactivism attempts to transcend the out-there/in-here dichotomy by focusing on the *interaction* between a person and his or her environment. This theory suggests that the individuals and the environment are mutually responsive: As individuals, we do not simply react to a static environment around us, nor are we isolated, contained individuals who manipulate our surroundings. Instead, reality is *brought forth* on a moment to moment basis through our actions and interactions with others and with the environment.

The cognitivist/constructivist/enactivist discussion above has parallels with the three "verses" discussed in the previous chapter. A *conversa*, which focuses on a reality brought forth as persons attempt to understand an other's experiences, explanations and mathematical distinctions, will be used throughout this work to represent an enactivist position on knowing and reality. From this perspective, "truth is not born nor is it to be found inside the head of an individual person, it is born *between people* collectively searching for truth, in the process of their dialogic interaction" (Bakhtin 1984, 110). In enactivism, objectivity/subjectivity of knowing and experience is not an either/or, but a both/and phenomenon. Philosophers have recently tried to transcend the dichotomy that has been created between objectivity and subjectivity by proposing that subject and object are interconnected. In mathematics,

> the individual is always part of what he or she sees. For example, in handling some practical mathematics apparatus I am finding out about myself. The apparatus is only meaningful insofar as it resists and guides my actions. The apparatus and my body become unified in any action. (Brown 1994, 91)

Enactivism suggests that knowledge is not stored in any location; rather, it only arises in action. We exist in relationship, in conversation with our surroundings which includes both the physical environment and the people in our lives. A friend and colleague of mine, Dr. Rebecca Luce-Kapler (1996), provided an enactivist image of knowing through the telling of a story. The following is a segment of that story:

> After my grandmother died, my mother gave me Grandma's two quart button jar filled with all the buttons she had collected over the years. Spilling those buttons

out onto the table, I can still find the large white button tied with store string that we used as a hummer toy. There is the green cloth button from her suit that she made for my uncle's wedding. The green suit I threw up on after getting carsick in my grandfather's brand new 1963 red Strato Chief. . . . What these buttons have taught me is that the memories of my grandparents are not inside me, preserved in some little jar of their own, but rather exist in my choosing and touching of the buttons, which seem to call up memory. The boundary between the buttons in the jar and my memories is not clear. Memory is neither in me nor in the buttons, but in our interaction, in our conversation. (1–2)

Enactivism attempts to blur the distinction between knowledge and action as was suggested in the above story. In this view, language and action are not observed as outward manifestations of some inner mental functioning; rather, they occur in interaction and are the visible aspects of a person's embodied or enacted understandings (Davis 1995b). The aphorism used in enactivism, "knowing is doing" (Maturana and Varela 1987), suggests that a person's actions are not a deliberate execution of some mentally predetermined sequence, nor are one's actions completely random. At every moment the environment is open to possible actions while restricting the possibility of other actions. Similarly, a person's phenomenological history opens possibilities for some actions and restricts others. The sphere of possibilities is shaped by the person's biological and experiential history of interaction and by the physical and social environment in which the person participates. As such, mathematical understanding is not observed as a knowing "that" or a knowing "how"; instead understanding is observed as implicit in a person's spontaneous actions and interactions with others with and in an environment (Taylor 1991; Shotter 1993). A focus on knowing as it occurs in interaction accounts for a physical environment, the phenomenological histories of the persons, as well as the history and culture in which the conversation is embedded; all of these aspects are enacted on a moment-to-moment basis.

Enactivism emphasizes the relationship between our biological structures and an environmental medium. Our biological makeup is seen as what simultaneously separates us and places us in relationship with others and with our environment. So even further than overcoming a mind/body dichotomy, enactivism stresses the symbiotic relationship between us and the reality we bring forth; or as Sumara (1996) writes, the ecological relationship of "us/not-us." Knowledge is contained neither within us nor in the world, but through the interactions of us/not-us. Such a relationship suggests that the/a world does not have predetermined features that we somehow react to or manipulate. Instead, reality is *brought forth* in these us/not-us interactions through a process of cognition. Cognition, from an enactivist perspective, is said to be the continual bringing forth of a world in the process of living.

Similarly, Gadamer (1989) insists that cognition or understanding is not a method but an event. Knowledge is not something one can appropriate and utilize within a specified time period, but rather it is an event enacted in the moment over which one has only partial control. "Every cognitive act takes place at a point of intersection of innumerable relations, events, circumstances, and histories that make the knower and the known what they are at that time"(Code 1991, 269). Understanding and reality do not exist prior to the interaction between the knower and known, but rather come into existence as an event of their interaction.

To further reveal the relationship between understanding, embodiment, and reality, Johnson (1987) writes that,

> Understanding is the way we "have a world," the way we experience our world as a comprehensible reality. Such understanding, therefore, involves our whole being—our bodily capacities and skills, our values, our moods and attitudes, our entire cultural tradition, the way in which we are bound up with a linguistic community, our aesthetic sensibilities, and so forth. In short, our understanding is our mode of "being in the world". (102)

Enactivism views cognition as the ongoing process of bringing forth a world which is bounded by what both the environment and the person's phenomenological history will allow. Therefore, activity is not viewed as proceeding through pre-established questions or procedures or as a linear series of problems and obstacles to overcome; instead, similar to a hermeneutic research approach, ideas, questions and suggestions present themselves to the learners and it is by acting on these ideas that the participants' reality is brought forth and their knowing is observed to occur. Varela, Thompson and Rosch (1991) suggest that,

> the usual tendency is to continue to treat cognition as problem solving in some pregiven task domain. [However], the greatest ability of living cognition . . . consists in being able to pose, within broad constraints, the relevant issues that need to be addressed at each moment. These issues and concerns are not pregiven but are *enacted* from a background of action, where what counts as relevant is contextually determined by our common sense. (145)

It was emphasized in the previous chapter that the practice of mathematics when viewed from a broad and historical perspective is not an action on a preformulated problem, but rather, mathematical practice arises from questions and concerns as a mathematician interacts with a mathematical phenomenon. It is important to note that the mathematical phenomenon is also not an entity that preexists action, but is brought forth by a mathematician through a multitude of ancestral and lived physical and interpersonal interactions. From this perspective, problems are formulated and solutions are reached; however, they are simply part of the ongoing cognitive process of living, rather than beginning and ending points or obstacles to eliminate. In the previous chapter, I described mathematical practice in terms of problem solving and suggested the difficulties and perhaps the narrowness of such a perspective. Enactivism provides a means for us to view the practice of mathematics specifically and intelligent behavior in general not as success in discrete problem-solving activities but as a continual process of bringing forth a world through the ideas and questions that arise in our us/not-us interactions, in our conversations with others and otherness.

Within an enactivist perspective mathematics can be seen as an innate part of our cognitive/bodily makeup. The body of mathematics is not a collection of detached truths contained either inside or outside of us, nor is it a process derived from proof or problem solving. Learning mathematics affects who we are, what we do, how we stand in relationship to others, and how we situate ourselves in our world. Mathematics contributes to our perceptions and bounds our actions. It helps shape the world we bring forth and within which we act (Davis 1995b).

Our knowledge and experience shape, even constrain, our actions and perceptions, and, hence, the reality we bring forth is similarly shaped and constrained. Gadamer (1989) uses the term "prejudice" to refer to the knowledge and experiences that shape our understandings and perceptual capacities. Our prejudices work not simply to limit our perceptions and constrain our actions, but are in fact what make perception and action possible. Without them we would be unable to draw meaningful images out of the noise of perceptual possibilities that surround us. How we see the world or what world we bring forth is in part shaped by the mathematical perceptions we have developed from a long human history of mathematical developments such as base-ten counting methods, Euclidean geometry of lines and circles, statistical and probability references to everyday events, and fractal representations in nature. When appropriate for the moment, we choose to see, analyze, and explain our experiences through mathematical eyes. Our mathematics is and becomes an integral part of who we are. It permeates our culture and frames our thinking. While our mathematics may narrow our perception in some ways it also links us to a world of rhythm and pattern. Viewing mathematics as the practice of problem solving—when problem solving is described as a source of practical applications, as an attempt to match the world, or as obstacles to overcome—is a particular "prejudice" of the practice of mathematics.

Although we can never become fully aware of our assumptions or prejudices it is important to acknowledge their existence and recognize the necessary constraints they place on our knowing and doing. Through our prejudices we are able to select or make distinctions on ideas and objects, but for every distinction made in one way, we close off possibilities for other distinctions to be made. Recognizing that our perceptions cannot be judged in terms of "correctness" or "certainty," but only in terms of meaningfulness and coherence to experience, we become open to and seek other possibilities for knowing. It is when we accept our perceptions as certain, that we stop asking questions (Maturana 1998). A recognition of prejudices allows one to put aside the arrogance of believing in the correctness, certainty, or completeness of his or her own explanations of lived experiences and accept a more humble position. The act of opening oneself to alternative possibilities involves becoming sensitive to the other's alterity, for it is only through being open to the other that we are forced to question and possibly revise our assumptions. Conversation provides an opportunity for such openness and otherness.

In a conversarial reality interacting with others is not simply helpful in clarifying our own explanations; rather, the possibilities for acting, explaining and understanding suddenly expand when there are opportunities for interaction. It is by participating in conversation and listening to the explanations of others that we can question our own assumptions, expand our opportunities for experience, and broaden our understanding. Human existence takes place in the relational space of conversation (Maturana 1988). Our bodies act as points of intersection in an evolving network of us/not-us interactions. Conversations are the flow of language, action and emotion. Our world, our reality is brought forth through our conversations with others.

Learning mathematics in school is essential, but not only for narrow utilitarian reasons such as providing students with problem-solving skills necessary for adult life. We also need to study mathematics to understand our own prejudices and to explore other possibilities for acting and knowing. To deny the study of mathematics lessens one's experience of living.

Language in Interaction
A view of conversation which emphasizes embodied relationships and interactions requires a broader view of language. The emphasis on communication in mathematics education has brought attention to language as it is used in learning mathematics. Research in language use,

such as discourse analysis, often focuses on the words actually spoken, but in conversation we know that much is said or meant without words: "To hermeneutically understand language is precisely to pierce through the facade of uttered language in order to bring attention to the things which our words attempt to share, but without fully succeeding, hence their indigence or 'secondarity'" (Grondin 1995, 143). Conversations cannot be confined to words. In our everyday conversations with others we take into account that which was said and done, what was previously said and done, and that which one was not able to say and do. "Speech and understanding thus emerge as 'speculative' processes whose success is nothing less than fragile" (152). The lack of clarity and certainty in our expressions is not a limitation of language or of ourselves as humans. The incompleteness of expression is a demonstration of the ongoing incompleteness of our existence and meaning-making activities (Maturana 1998).

We recognize that conversation is more than simply the words spoken. Yet, we often refer to the words and the actions in conversation as separate and distinct processes. Abram (1996) suggests an alternative which describes both words and actions as embodied gestures. He views the genesis of language as a phenomenon of both mind and body arising as a physical gesture, as the "bodying-forth" of meaning and emotion:

> Active, living speech is just such a gesture, a vocal gesticulation wherein the meaning is inseparable from the sound, the shape, and the rhythm of the words. Communicative meaning is always, in its depths, affective; it remains rooted in the sensual dimension of experience, born of the body's native capacity to resonate with other bodies and with the landscape as a whole. (74–75)

Bakhtin's (1986) notions of language also emphasize the contextual aspects of language and the interconnectedness between ourselves and others. He reminds us that our words and actions are not wholly our own. "Our speech, that is, all our utterances (including creative works), is filled with others' words, varying degrees of otherness or varying degrees of 'our-own-ness,' varying degrees of awareness and detachment" (89). Although education has emphasized the importance of creating autonomous learners, I question whether we, as human beings, are ever capable of being autonomous, and I wonder what is "closed off" by making such a distinction. Although we tend to think of ourselves as independently constituted, our identities are tied closely to the identities of those around us. Words and actions cannot be viewed as a succession of disjointed autonomous products, but rather are inherently responsive to a chain of other events, including past utterances and actions, and the anticipation of future utterances and actions. A person's words and actions carry assumptions and implications tied to the moment that they occur, which in turn create a new context and potentially new possibilities for the actions of the other.

Although there are often attempts in research to segment out one person's ideas from another's, in conversation these ideas cannot be understood out of their interactive context. However, this context has multiple layers. Actions and utterances draw upon interactions that the person experiences within the immediate context, upon a history of interaction brought from previous experiences, and also from an embodied ancestral history brought forth to that moment. Our choice of words, the unique utterance created, is "shaped and developed in continuous and constant interaction with others' individual utterances" (Bakhtin 1986, 89).

> Utterances are not indifferent to one another, and are not self-sufficient; they are aware of and mutually reflect one another. These mutual reflections determine their character. Each utterance is filled with echoes and reverberations of other utterances to which it is related by the communality of the sphere of speech communication. (91)

A person's activities are never wholly his or her own—they are "filled with echoes" of the actions and utterances of others.

Bakhtin asserts that all communication is interactive or dialogical; however, the quality of its "addressivity" changes as the speech genre of the communication changes. A speech genre, according to Bakhtin (1986), is a set of "relatively stable thematic, compositional, and stylistic types of utterances" (64). From the first words, we guess at and cast our speech in the genre of our interactions. Rather than a focus on the speaker, who does not necessarily have any relation to the listeners, a speech genre takes the role of the other into account, thus highlighting the nature and type of their relationship and the topic of their interaction. "Each speech genre in each area of speech communication has its own typical conception of the addressee, and this defines it as a genre" (95). Here, addressivity, or the quality of turning to the other, raises an important distinction between mathematical conversation, as a potential form of speech genre, and other forms of interaction or discourse. Conversation has the quality of "colloquium" or interacting *with* the other, while other forms of academic talk, such as argument, often have a quality of "alloquium" or talking *to* the other (Simpson 1995). The addressivity in mathematical conversations implies a responsiveness and a personal responsibility to the other. One listens to understand, and one speaks to share experiences and understanding. There is an ethical respect implied in our conversational interactions with others. We guide what we want to and can say in relation to how we feel these others will respond (Shotter 1993).

In many respects, addressivity as the quality of turning (versing) to or with another may be indicative of the reality or world (versum) brought forth. If in validating my mathematical explanation, I bring forth a *universe* in which only one explanation is certain, I *turn to* the other to compel him or her to see and agree with that truth. However, if I bring forth a *conversarial* reality, many explanations are possible, but together we strive to come to a mutual understanding of an experience or explanation by *turning with* one another in conversation.

Occasions for disagreement and argument are present in conversation, but the tone, structure and addressivity of disagreement are different.

> Knowers present their evidence and construct understandings through contextual and open-ended narratives in which analytic distinction, deductive argument, and replicable experiment may figure but do not predominate. Knowers take disagreement as an occasion for collaborative deliberation and communication rather than for debate. (Ruddick 1996, 262)

Gadamer (1989) writes that in "a conversation, when we have discovered the other person's standpoint and horizon, his ideas become intelligible without our necessarily having to agree with him" (303). Agreement and disagreement are aspects of conversation, but they are not of primary focus. Rather, the purpose for participating in conversation changes to the sharing of explanations as part of one's ongoing efforts to make sense of his or her experiences in coexistence with others. Therefore, "to reach an understanding in a dialogue is not merely a matter of putting oneself forward and successfully asserting one's own point of view, but being transformed into a communion in which we do not remain what we were" (379).

While I have attempted to provide a philosophical basis for conversation by describing its qualities in terms of cognition, interaction, and language, what forms might cognition, interaction, and language take in mathematical practice if a space were created for conversation? What are the qualities and features of a *mathematical* conversation? While the present chapter hopefully provided a general understanding of conversation as a particular form of interaction, the linear nature of this research does not reveal the all-at-once quality whereby my observations of mathematical interactions also shaped the understanding of conversation presented thus far. The following three chapters are an exploration of mathematical conversation through the use of illustrative examples. They simultaneously elaborate on the qualities of cognition, interaction, and language presented in this chapter, but also point to other features and qualities unique to mathematical conversations.

Chapter 5

Explanations in the Process of Understanding

We live our lives as curious, passionate beings. The happenings of life are encountered as if out of nowhere—as moment-to-moment experiences in our lived histories (Maturana 1988). As curious beings we ask questions about our experiences which lead to attempts to try to explain, and understand our lives in interaction with others. Many of these questions arise from our mathematical prejudices and perceptions of the world; therefore, our attempts to explain our experiences frequently draw on our mathematical understandings. Experience "is to be understood in a very rich, broad sense as including basic perceptual, motor-program, emotional, historical, social, and linguistic dimensions . . . experience involves everything that makes us human—our bodily, social, and linguistic dimensions" (Johnson 1987, xvi). In a conversarial reality our explanations are not of an independent world, but rather, explain our own experiences in coexistence with others. The explanations we accept give rise to new experiences, new questions, and new explanations in a recursive cycle. Our experiences, our conversations with and in the world, and the subsequent explanations are and become our reality.

In the present reform movement in mathematics education, the responsibility for providing mathematical explanations for specified problems is shifting from the teacher to the students. Students are given a space to voice their arguments for and against their own and their peers' solutions to tasks and problems posed by the teacher and occasionally by students. The explanations provided by students afford a window into student reasoning that was not available when they provided only one word or numerical answers to teacher questions.

What questions and subsequent explanations would arise if mathematical problems were not prespecified with foreseeable end goals? This chapter examines explanations as they occur within a mathematical conversation.

How are explanations posed within a conversational context? What reality is brought forth? What explanations are accepted as mathematical and how is this acceptance signaled? These are a few of the questions that arose through the following research experience.

Introducing the Participants and the Prompt[1]

At the time this research began, Tamera had completed her Bachelor of Science/Bachelor of Education combined degree and Kylie had completed a Bachelor of Science specializing in mathematics, as well as her first year of a two-year Bachelor of Education after-degree program. Throughout their postsecondary education, Tamera and Kylie had each completed a number of advanced courses in university-level mathematics, including several calculus, linear algebra, abstract algebra and (primarily Euclidean) geometry courses. Their backgrounds indicate that they have both been "successful" in mathematics in the past. Success is broadly defined: they were interested in and capable of pursuing mathematics as careers and had successfully completed the required courses. Rather than presenting this research as how people *should* come to know mathematics through interaction, this chapter suggests how some people, at least, *do* come to know mathematics through mathematical acts of explanation and reasoning within a context of conversation. Kylie and Tamera's investigation was chosen to illustrate the interactions of two relatively successful mathematics students engaged in a mathematical conversation, the questions that arose, and the subsequent explanations that occurred as a result of their interaction.

This was Kylie and Tamera's fourth session with me. In previous sessions they had investigated consecutive sums, a microcomputer geo-world, and three-dimensional objects with two-dimensional representations. They, along with all other pairs in this study, were given a mathematical prompt to begin the session. On this occasion, the Diagonal Intruder prompt, as shown in Figure 1, was provided (adapted from Stevenson 1992).

Interpretations made of Kylie and Tamera's mathematical explanations as they occur within their conversation are embedded within a framework of *enactivist* inquiry. Enactivism suggests that mathematical understanding should be studied not through its products or mental structures in and of themselves, but rather as understanding occurs in the interaction between persons and their environment in the process of bringing forth a world of mathematical significance (Maturana and Varela 1987; Varela, Thompson, and Rosch 1991). The direction of their/our conversation and the mathematical ideas investigated were occasioned by their interac-

Diagonal Intruder

An intruder keeps breaking into the rectangular shaped hotels. He enters at one corner and exits out the opposite corner. The rectangular shaped hotel owners are worried and want to know how many rooms the intruder will invade in their hotels.

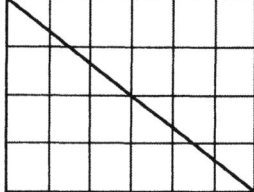

This is an example of a 4 x 6 rectangular hotel.
How many rooms does he enter here?

Figure 1. Diagonal intruder

tions with each other, myself, the prompt, and the materials that were available to them, such as the graph paper, ruler, and the multiple pieces of paper with mathematical diagrams and symbols that they produced throughout this one-and-a-half hour session. A person's activities—conjecture making, record keeping, object manipulation—arise through interactions with others and with the environment. This is not to say that the other person, or the problem, or the manipulatives available are able to project themselves into students' thoughts to give them instructions or ideas. Rather, these interactions act as occasions in which the individuals, in their own way, include the other, the problem, or the physical environment in their knowing and in their actions. Thus, a person's knowing/actions are not independent of the other, and in the same way the knowing and actions are not independent of the place or of the moment in time. All of these forms of "us/not-us" interactions (Sumara 1996), which are and become the artifacts of their mathematical practice, provided opportunities for Tamera and Kylie to generate questions and subsequent explanations of their experiences.

Narrative: Generating Examples and Local Number Patterns

Tamera and Kylie were given the Diagonal Intruder prompt on a sheet of paper. After reading the prompt Kylie initiated a course of action. "Well

Figure 2. Rectangles with widths of 1

should we start off by going. . .1 by 1?" She outlined a small square on graph paper with a diagonal sketched through it (see Figure 2).

Kylie continued, "1 by 1. Obviously it goes through everything. 2 by 1," drawing a 2 x 1 rectangle and diagonal as Tamera watched. "So if it is 1 row or 1 column, it's going to go through everything."

"Uh huh," Tamera agreed.

Kylie recorded the conjecture at the top of her page: "1 row or 1 column—goes thru all."

"Two by two?" Kylie suggested the next example and then made a conjecture after drawing it: "So if it's a square it's going to go through everything too."

"No—" Tamera disagreed.

"Or through the diagonal anyway."

"Yah. Through half of them."

Kylie and Tamera discussed how to write this and came up with "square—goes thru # of the dimension."

After stating their conjecture for squares they moved on to rectangles with a constant width of 2. Kylie began to systematically draw rectangles from 2 x 1 to 2 x 8. When she got to a 2 x 3, Tamera got a sheet of graph paper and started to make her own records. Although both Tamera and Kylie made the same diagrams they checked after each one to make sure they both got the same number of "rooms entered." Kylie's records show eight rectangles (see Figure 3). Each rectangle is labeled with its dimensions as well as the number of squares the diagonal crosses through (or the number of rooms entered by the Diagonal Intruder). After eight rectangles were drawn, Kylie and Tamera both stopped to look at the number pattern created and to make conjectures about its regularity. The following portion of their investigation was their attempt to *explain* the number pattern they saw for the number of rooms entered in hotels with a constant dimension of 2.

Kylie began with a conjecture; "I think for this one, on the evens they always go through—" (see Figure 3).

"—the center," Tamera pointed to the diagonal's intersection point in the middle of the rectangle.

"—the middle you know? Like on all the even ones. So that's how come it doesn't go up. Like it stayed at 6." Kylie noticed that in the

Explanations in the Process of Understanding 61

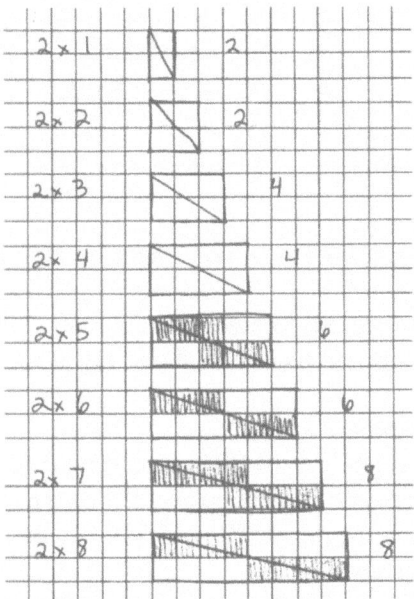

Figure 3. Rectangles with widths of 2

rectangles 2 x 5 and 2 x 6 the number of rooms entered remained at 6. Although it was never discussed or negotiated, Kylie and Tamera acted under the assumption or rule that whenever a diagonal crossed through an intersection point on the grid it did not enter all four rooms adjacent to the intersection, only the two rooms where the diagonal entered and left.

Tamera agreed, "Right, 'cause it just goes through the center."

"It just goes through the middle so it isn't going through any more. But then when we add another one, the next odd one, then it goes through two more because you have to add the middle ones."

"Yah."

They looked at their diagrams for several seconds in silence. Finally Kylie looked up and asked me, "Do you want us to write that mathematically?"

I responded, "Sure. See if you can do that," although at this point I was uncertain how the mathematical notation would or should look.

Still thinking about their conjecture, Kylie said, "It goes through 2. So our even ones go through a middle. I don't know how we write that." Tamera did not respond. They looked at their papers in silence for almost a minute.

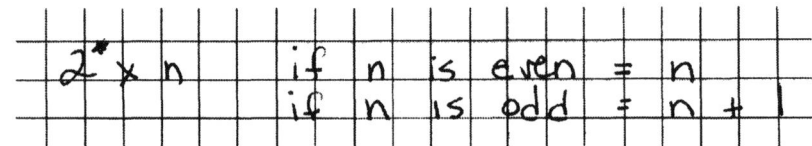

Figure 4. If-then condition statements for 2 x n rectangles

I asked, "Could you write how you might be able to predict a 2 by anything? Like a 2 by *n*?"

Kylie looked at the pattern formed by the 2 x *n*s. "Hmmm. If you do—it depends on if it is odd or even. Like if you say a 2 x 20, I can say it is going to be 20."

Tamera added, "And a 2 x 19—"

"And 2 x 19 is also going to be 20 because you'd go down one." Kylie paused to record their ideas, "So if you go 2^n, then if *n* is even—Whoops," Kylie scribbled out the exponent, "2 x *n* it's *n*." Kylie stated the "evens" aspect of their conjecture.

Tamera agreed, "It equals *n*," and added the odd number aspect, "And if *n* is odd it's *n* + 1."

Kylie recorded these conjectures (see Figure 4). They looked at these mathematical statements for a moment and then Kylie looked at me, "That's the only way I can think of."

I responded, "That's good," and all three of us nodded our heads in agreement.

Tamera proposed the next course of action; "Let's go on to 3s."

Discussion of Examples and Patterns Narrative

The Prompt

The Diagonal Intruder lends itself to diverse and undetermined paths of exploration. It does not funnel participants towards a particular set of relationships. Therefore, the conjectures and explanations made by Kylie and Tamera were dependent on the mathematical skills, interests, and experiences that they brought to the conversation. Tamera and Kylie seemed to share an inclination towards identifying and explaining number patterns in the mathematical prompts that were provided in this and in previous sessions. That is, they consistently attempted to bring forth an image of number patterns and used that aspect of the body of mathematics to make sense of their experiences. The Diagonal Intruder, however, did not *instruct* them to do so. While the Diagonal Intruder allows for these images, different questions may have given rise to explorations utilizing different networks within the body of mathematics. After their investigation, I shared with Kylie and Tamera that other participants attempted to come to an understanding of the Diagonal Intruder while basing their image of the prompt, appropriately or not, on number patterns, the Pythagorean theorem, trigonometry or geometry. Both Kylie and Tamera were surprised and indicated that they were unaware that other perspectives or mathematical acts were possible.[2] While other paths of mathematical activity were possible, Kylie and Tamera's explanations explained the number pattern world they brought forth in interaction with each other and with the Diagonal Intruder.

Explanations for 2 x ns

The expectations Tamera and Kylie brought to this investigation and to several previous investigations were to distinguish and bring forth number patterns in their actions. The pattern 2, 2, 4, 4, 6, 6, and so on is not thought to preexist their actions, but is brought forth through their actions and the distinctions they made. In relation to the "prejudice" towards number patterns, Kylie and Tamera also distinguished "odd" and "even" aspects of the pattern to explain their actions. After discussing their observations, Kylie offered the following explanation: "It just goes through the middle so it isn't going through any more. But then when we add another one, the next odd one, then it goes through two more because you have to add the middle ones." This initial attempt does indeed explain the phenomenon, but their explanation was not accepted by them or myself, the present community, as a mathematical explanation. Only when their explanation was written using mathematical symbols and was capable of predicting any 2 x n was it accepted as a mathematical explanation. The expectation arising from this and from previous experiences was that explanations should be "written mathematically" in some form. My question, "Could. . .you. . .predict. . .?"—occurring after a period of silence—was not an attempt to instruct them to write their findings in a particular way, as I myself did not know how the findings might be written, but it was a question arising from their actions and record keeping. In fact, I mentioned the notion of "predicting" several times throughout their investigation, and it likely reflects what *I* consider to be an effective and legitimate mathematical action.

After writing their if-then condition statements for the 2 x ns (see Figure 4), Kylie *offered* it to us as the/an explanation: "That's the only way I can think of." An identifying feature of this and other mathematical explanations in conversation is that explanations are offered to oneself and to the other as a way of making sense of their experiences together to that moment. We can imagine that Kylie could say, "That's the only way I can think of *for now*," implying that they could return to the explanation at anytime for revisions. Explanations offered in the course of conversation do not necessarily insist on being a "complete explanation" or the final word on the subject.

The explanation offered satisfied the present community's criteria for acceptability, which appeared for Kylie and Tamera to be an explanation which provided enough of an understanding of the mathematical situation to allow them to continue on. Explanations in a mathematical conversation do not attempt to satisfy a criterion of correctness or completeness, nor are they offered as a convincing argument. Acceptance of the if-then statements in Kylie and Tamera's activities was given by myself verbally, ("That's good") and by all of us physically by nodding our heads. My verbal acceptance of the explanation, "That's good," could have also been stated as, "That's *good enough for now,*" emphasizing the temporal nature of the explanation. Although the explanation was not a proof, or an explanation to convince, it was an explanation that was coherent with their interactions thus far. The explanation that was offered by Kylie and Tamera was accepted as plausible, sufficient, and believable for the present moment. More importantly, the explanation was not an endpoint for them. Once offered and accepted, it allowed them, as Tamera said, to "go on to 3s." Accepting the condition statements as an appropriate explanation for the number pattern arising through their interactions allowed Kylie and Tamera to *continue on* to the next set of rectangles. Such acceptance laid a path for future actions and explanations and defined a domain of legitimate actions, or legitimate mathematical practices, and acceptable explanations.

The rhythm of making predictions, offering conjectures, and providing possible explanations appears central to a conversational discourse. Acceptance of an explanation is not made through judgments of right or wrong, correct or incorrect. Instead, explanations at the time they are proposed are accepted if they are viable and informative given the possibilities and constraints of the immediate situation. The viability of the explanation has to appease both participants in the conversation. This is not to say that a compromise has been reached, but the explanation itself is a coemergent phenomenon arising from Kylie and Tamera's actions and interactions to that moment.

The coemergent explanation validated on their experiences indicates that a conversarial reality has been brought forth. Rather than being accepted as *the* explanation, the if-then condition statements were accepted as a *plausible* explanation. A plausible explanation is not a definitive solution; instead, it need only be "good enough." That is, the explanation is not necessarily ideal, but it is possible, it fulfills the present needs, and it provides a useful and viable structure for their actions (Varela, Thompson, and Rosch 1991). The plausible explanation became a means through which Tamera and Kylie could continue further into the investigation—with the possibility that they could return to it later to revise it if necessary. Explanations are significant points on the path of mathematical activity. They may be revisited and revised when the actions and assumptions upon which the explanations are based are later called into question.

Throughout this session and earlier sessions, Kylie and Tamera acted in expectation that they were the ones responsible for the path of mathematical activity. They knew that they were not expected to reach a particular answer or conclusion and that no "solution" would later be available to them or me to compare how successful their actions were or how far they progressed. The reality brought forth by Kylie and Tamera as they engaged in this mathematical activity was not directed by a question that was posed or imposed pristine and predetermined by the Diagonal Intruder prompt, but rather their interaction with each other, me, and the prompt raised questions begging explanations. The explanations offered were recognized as incomplete. Explanations that were accepted were perhaps one of many possible explanations, even though Tamera and Kylie may or may not have considered what those alternatives might be.

Explanations that are accepted in a mathematical conversation are not endpoints or conclusions, but rather are attempts to understand the questions that the participants pose themselves. Explanations are viewed as acceptable, or good enough, when they broaden the participants' understanding of a phenomenon brought forth and allow them to maintain their interactions with each other and with the mathematical phenomenon. The explanations offered and accepted in conversation are formative and meaningful, and act as opportunities for participants to continue on to new experiences, new questions, and new explanations.

Narrative: Where Do the Dots Go?
While Tamera and Kylie brought forth an image of the Diagonal Intruder as a number pattern exploration, many questions arose from their interactions and from the artifacts of their work. A recurring and *mutual concern* raised by Tamera and Kylie was determining whether or not they drew their diagonals accurately. An inaccurate drawing could lead to an inaccurate count of rooms entered, which would affect their number pattern

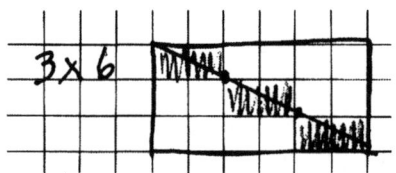

Figure 5. Kylie's 3 x 6 rectangle

conjectures. The following narrative illustrates their concern for making accurate drawings. This concern was raised throughout their investigation. It began early in their activities when they both drew 3 x 6 rectangles (see Figure 5) and resurfaced later when they drew rectangular sets with constant widths of 6 and 7 (see Figures 6–8 and 9–10).

Tamera looked at her drawing of a 3 x 6 rectangle and said, "It looks like it goes through this one," pointing to a possible intersection point, "but I'm not sure if it's supposed to. 7?"

Kylie looked at her own drawing, shown in Figure 5, and said, "Ah, yah."

Still unsure, Tamera asked, "6? Is that—"

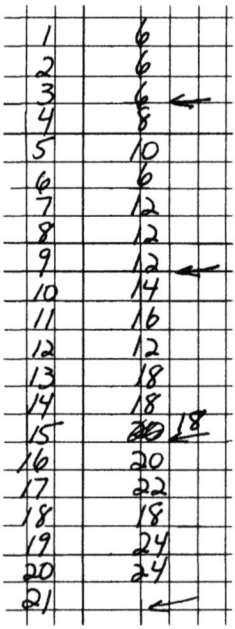

Figure 6. Number pattern for 6 x n rectangles

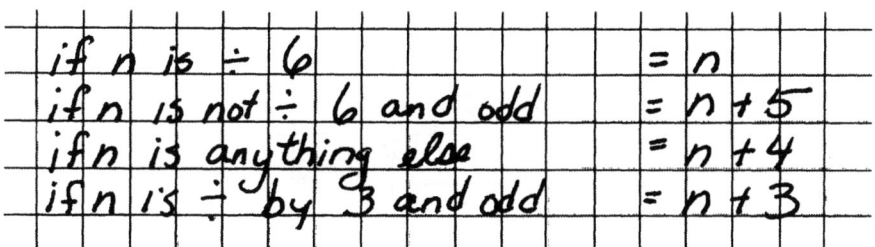

Figure 7. If-then conditions statements for 6 x n rectangles

Kylie counted again. "See maybe, maybe 7."

"Well I don't know if it goes through here," Tamera said, pointing to a possible intersection point. "Looks like it."

Kylie turned to look at Tamera's diagram, "Yours looks more exact than mine."

More certain of her answer Tamera said, "It should go right through those."

"Oh, yah. Well it has to go through those. So only 6?"

"Yah," Tamera agreed.

The concern raised for determining whether the diagonal was drawn accurately surfaced several times throughout their investigation. The first mention of it above, occurred within the first ten minutes of their investigation. The concern resurfaced approximately fifty minutes later when they tried to determine a number pattern (see Figure 6) and the subsequent if-then condition statements for widths of 6 (see Figure 7).

They originally devised a set of three conditions (the fourth shown in Figure 7 was added later). As they checked their conditions, they noticed that the number pattern arising from the rectangles with widths of 6, shown in Figure 6, repeated in groups of 6, starting and ended with a multiple of 6:

6, 6, 6, 8, 10, 6;
12, 12, 12, 14, 16, 12.

However, their next set of six did not seem to follow the same pattern:

18, 18, 20, 20, 22, 18.

A consistent pattern would have read 18, 18, 18, 20, 22, 18. The first "20" in the list above is inconsistent with the previous patterns. Although

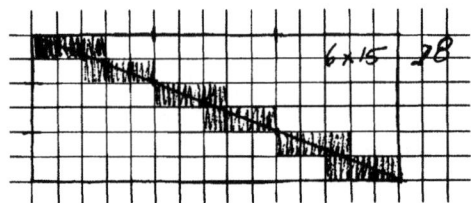

Figure 8. 6 x 15 rectangle

Tamera and Kylie eventually changed the 20 to an 18, the narrative continues as they attempted to determine what had gone wrong.

Kylie looked at the numbers generated in Figure 6. The numbers on the left indicate the length of the rectangle and the numbers on the right indicate the number of rooms entered or squares the diagonal crossed through. "So our problem is in between [6 x 14 and 6 x 16]," referring to the inconsistent "20" written for the 6 x 15 rectangle. "So 15," the 6 x 15, "should really be 18."

"It should be according to the pattern."

"But according to this," referring to their condition statements produced to that point, "it works." That is, a 6 x 15 rectangle falls under condition statement (2): if n is not $\div$ by 6 and odd = $n + 5$. Since 15 is not divisible by 6 and it is odd, the number of rooms entered should be 15 + 5 = 20.

I interjected, "Do you want to check your picture again?"

Kylie asked, "The picture of—?"

"6 x 15," where the problem appeared to lie (see Figure 8).

"Oh maybe I screwed up. It could go through right there I guess" referring to a possible intersection point on her diagram.

"That's a good question," I mentioned. "How do you decide where those dots go?" referring to Tamera's practice of noting intersection points with large dots. "Is there a way you could—?"

Still looking at her diagram Kylie continued, "And right there. That would take off 2. But they—"

Tamera said, "Then that would make it fit" with their number pattern.

Kylie agreed, "Uh hmm."

"Then maybe if it is—"

"Then this would be 18," Kylie referred to the 6 x 15 part of the pattern, "16 would be 20, 17 would be 22 and 18 would be 18."

Now that the pattern was consistent, Kylie returned to the question I posed earlier. She restated it, "Where do you decide where those dots go?" There were thirty seconds of silence before Kylie attempted to explain. "If the number is divisible by—both." Another 30 second pause took place as both Kylie and Tamera studied their diagrams. Kylie hesitated but continued to offer the beginnings of an explanation, "I don't know, but look, both numbers here [length and width] are divisible by 3."

Tamera echoed, "3."

Kylie continued, "And so is this one," referring to the 6 x 9 rectangle. "6 is divisible by 3 so you get 3 sections of 2. And here," pointing to the 6 x 15 rectangle, "you get 3 sections of 5." Kylie placed hash marks on the sections she was referring to (see Figure 8). "It's where it matches up. You were marking those," referring to the dots that Tamera made on her diagrams. Kylie turned to Tamera.

"Yah, it works. Okay these are both divisible by 6," in the 6 x 12 rectangle, "so you get 6 sections there."

"And 2 sections there. Yah, so you *can* figure out where they are going to go. Cool."

They realized that they needed a fourth condition to account for some of the numbers not fitting into their pattern. Prior to changing their response to the 6 x 15 rectangle from 20 to 18, they had placed arrows towards the numbers 3, 9 and 21 which did not fit into their pattern (see Figure 6). They originally tried to formulate a fourth condition based on powers of 3, but were unsuccessful. Tamera eventually said, "We just have to say if n is whatever, we have to think of a way to add—"

"How about if n is divisible by 3 and odd, both of those same things, then it's $n + 3$?"

Satisfied with this fourth condition, Kylie and Tamera completed the rectangles with a constant width of 6 and continued their investigation into rectangles with widths of 7. They now had a means for determining the intersection points. As they developed their if-then conditions for widths of 7, Kylie used their explanation for "where the dots go" to determine intersections within a 7 x 12 rectangle.

Starting fresh with a new piece of grid paper Kylie began to write the condition statements for rectangles with widths of 7 (see Figure 9). "So 7. It should be—if it is divisible by 7 it's n and anything else should be plus 6? 'Cause this is plus 5," referring to condition statement (2) for rectangles with widths of 6 (see Figure 7).

"Yah."

"Should we test this one? What did I say? If n is divisible by 7 then it's n, and if anything else it then should be plus 6. Plus 6?"

"Yah."

Figure 9. If-then condition statements for 7 x n rectangles

"How about—I'm going to try 7 x 12."

"I'm going to try 7 x 14."

Kylie acknowledged the choice, "14. It should—What was I doing again? I was doing 12 right?"

While Kylie was still thinking Tamera completed her example, "Well, this one works." Tamera counted 14 rooms entered (see Figure 10) which fit with their initial conjecture that if *n* is divisible by 7 the number of rooms entered is *n*. In this case, *n*, the length of the rectangle, was 14 and the number of rooms entered was 14 confirming their condition statement.

"Good. What am I doing? 7 and 12. They don't have any so they shouldn't cross." She completed her drawing and then counted, "18." She checked her answer with the conditions, "12 plus 6 is 18."

"That works."

"It's good to know where those dots will be—"

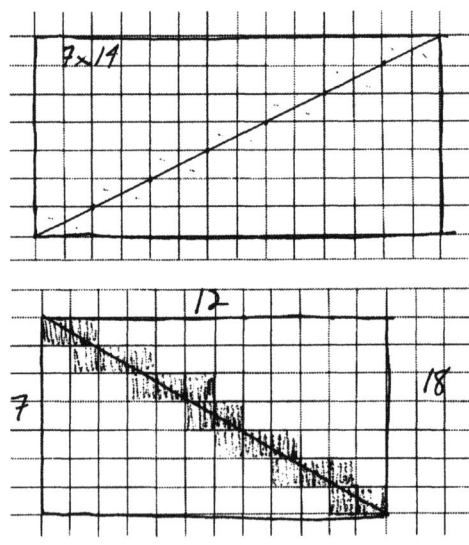

Figure 10. Tamera's 7 x 14 rectangle and Kylie's 7 x 12 rectangle

"Yah."

"—where those dots will be because otherwise, well you know when you are not being totally exact with your fat pencil."

Discussion of Where the Dots Go Narrative

Revisiting and Revising Explanations Stated in Action

Much of the focus of Kylie and Tamera's investigation was to find number patterns within sets of rectangles. Having the correct number of rooms entered according to the logic of their investigation was essential for noticing such patterns; thus, it is not surprising that an important concern stemming from their actions was trying to determine where the diagonal line intersected the horizontal and vertical grid lines simultaneously. As previously mentioned, Tamera and Kylie were acting on the shared assumption or rule that at points of intersection the diagonal only entered 2 rooms, rather than all 4 squares adjacent to that point. Tamera was uncertain whether the diagonal "goes through this one" when she looked at her drawing of a 3 x 6 rectangle, meaning she was not sure whether the diagonal intersected the horizontal and diagonal grid lines simultaneously at one point. This area of concern led Tamera to begin the practice of identifying where she thought the diagonals and grid lines intersected by indicating them with large dots. Kylie also began using this strategy in addition to her counting method which involved shading in squares the diagonal cut through.[3]

My question, "How do you decide where those dots go?" arose from their actions and our conversation. It was a question requiring explanation. The question flowed from Tamera's practice of drawing dots at points of intersection. The explanation they provided was tentatively based on divisibility—when both numbers share a common factor—although they never defined it as such. The need to find an explanation arose from our interactions with each other and the mathematical phenomenon. Their purpose for finding an explanation was not for justification, but to attempt to understand the phenomenon more thoroughly for themselves.

Tamera and Kylie both contributed to the explanation of "where the dots go." The explanation given was, "If the number is divisible by—both," if the length and width have a common factor, the rectangle can be divided into "sections" along its width and length. Where these sections "match up" is where the dots go. Although this explanation was vaguely stated, it was, "good enough" for their immediate purposes and it was understood and validated using other examples. The explanation is viable and informative and allows them to continue further into the investigation. It was an explanation stated *in action* and *in the midst of understanding* that allowed them to continue on. Once they became aware of "where the dots go" their experience with the Diagonal Intruder perceptually and cognitively changed. A different aspect of the body of mathematics was pulled into action involving common factors that provided a broader and different understanding of the phenomenon.

The explanation for where the dots go was accepted by both Kylie and Tamera. One signal of acceptance was that they both said, "Yah, it works." However, further and more substantial evidence is that their explanation became part of their tool-kit for analyzing subsequent sets of rectangles. When they examined rectangles with widths of 7, Kylie said, "What am I doing? 7 and 12. They don't have *any* so they shouldn't cross." Here, we can assume she means they don't have any common factors, or, in her language, no numbers divide evenly into both 7 and 12; therefore, no intersection points, or dots will occur. The task of counting the number of rooms entered by the diagonal became much easier and more reliable.

Formalization and Abstraction

It is not likely that Tamera and Kylie's explanation for determining where the dots go would be accepted if posed to an audience wanting justification or a convincing argument, but it was one that was accepted as plausible within the present community. It addressed their mutual concern and answered the question that had been posed. The explanation allowed Kylie and Tamera to maintain their conversational relationship with each other and with the mathematics. As a mathematics educator I may be somewhat discouraged that Kylie and Tamera did not attempt to mathematically formalize their thinking; yet, the impetus to formalize perhaps also needs to arise as a question of mutual concern flowing from their experiences. Any explanation offered in a mathematical conversation, formal or otherwise, must be perceived as broadening the participants' understanding and allow them to continue the conversation. At this point in their investigation, Kylie and Tamera appeared to be in a process of gathering data, making tentative conjectures and testing them through the use of specific examples. The questions they asked arose from these activities and their explanations satisfy their current concerns. They formulated meaningful explanations to their questions without extensive use of mathematical formalization to make sense of their experiences to that point.

Questions arise from one's interactions. They are not found outside a person's history of experiences and understandings. As educators then, we cannot necessarily "give" problems to our students and "instruct" them to engage in formalization, nor can we provide or promote explanations which involve abstractions and expect that students will find them more acceptable or meaningful. Instead, as we have seen through Kylie and Tamera's activities, questions arise from a person's interactions with others and with the environment. The desire to make sense of one's experiences, in relation to the questions the person him- or herself poses, leads to attempts to explain. Teachers may provide occasions for experiences which may give rise to questions requiring formalization, such as questions of "why?" or "is that consistent for all?" rather than simply descriptions of "what?" or "how?" However, a consequence of the conversarial view presented here is that educators have to recognize that these questions may not always be taken up as topics of mutual concern for students. Also, the teacher must relinquish the belief that some explanations are somehow truer or progressively better than others. This belief implies that a person is capable of testing the truth of an explanation by matching it to an independent source. Instead, educators may provide occasions for experiences so that the questions arising from the students' actions may be or need to be explained utilizing formalization and abstraction to provide meaning and coherence to their experiences.

Explanations are accepted in conversation, as was illustrated in Kylie and Tamera's mathematical conversation, if they satisfy a condition or criterion of *understanding;* that is, for a mathematical explanation to be accepted in conversation, it must provide a means for understanding a concern that is presently being addressed in the interaction. If mathematical abstractions are offered, but do not provide this understanding, the explanation may not be viewed by the student as useful or acceptable. It is important to note that proofs and other convincing arguments are not "more" acceptable or "better" explanations; rather, proofs are mathematical explanations arising from a particular set of concerns commonly posed by mathematicians and the explanations are offered to meet the criteria for acceptance for a particular mathematical audience. A mathematical proof is offered to an audience who judges its acceptance based on perceived completeness, accuracy, and aesthetic aspects (Hersh 1993). In a mathematical conversation the acceptance of an explanation is not based on accuracy, or completeness, but on whether it expands a person's understanding of the phenomenon at that moment.

Rather than being restricted to providing suitable environments and observing over the shoulders of students, it is possible for the teacher to become a participant in the conversation.[4] Any participant in a mathematical conversation may raise questions as genuine concerns arising from the observed interactions. For example, two of the questions I raised as a periphery participant in conversation were, "Could you write how you might be able to predict a 2 by anything?" and "How do you decide where those dots go?" These questions were concerns arising from Tamera and Kylie's interactions. I could not force them to address these concerns, nor should I expect them to be addressed in a particular way. While a teacher's questions and comments may receive more consideration, a conversation is not conducted from within any individual. The participants, together, must have a desire to address the concern. It must be seen as a topic of mutual concern.

The latter portion of Kylie and Tamera's activities above suggest that their explanations have become more abstract, not because abstraction necessarily follows from their concrete activities, but rather, their continued experiences and explanations allowed them to think differently about the mathematical phenomenon which utilized other aspects of the body of mathematics. At the beginning of their investigation Kylie and Tamera used a large number of diagrams to identify patterns, then built condition statements from the patterns. The experiential source for validating the condition statements was their physical drawings and number patterns. As they continued to interact in this environment, the number of physical drawings they made for each set became fewer and fewer. By the time they investigated rectangles with widths of 7 (see Figure 9), their condition statements were created using patterns found not in specific diagrammatic examples, but in the patterns within the condition statements themselves. The source for validating their condition statements became the pattern across the sets. Figure 11 shows are their records for rectangles with constant widths from 1 to 7.[5] The division sign is their notation for "divisible by." For example, in 5 x n rectangles "if $n \div 5 = n$" is read "if n is divisible by 5, then the number of rooms entered is n."

The condition statements for 7 x n rectangles were created by noticing a consistent pattern across the previous statements. That is, Kylie and Tamera noticed that each set contained two common statements: First, if $n \div c = n$, where n is defined as the variable length and c is defined as the constant width. This was the first condition statement written for each set. As an example, in a 5 x 15 rectangle, since 15 is divisible by 5 the

Explanations in the Process of Understanding

Dimension	Condition Statements	
	If—	**—Then**
$1 \times n$		Goes thru all
Squares		Goes thru # of the dimension
$2 \times n$	if n is even	$= n$
	if n is odd	$= n + 1$
$3 \times n$	if $n \div 3$	$= n$
	if n is anything else	$= n + 2$
$4 \times n$	if $n \div 4$	$= n$
	if n is odd	$= n + 3$
	if n is anything else	$= n + 2$
$5 \times n$	if $n \div 5$	$= n$
	if n is anything else	$= n + 4$
$6 \times n$	if $n \div 6$	$= n$
	if $n \div 3$ and odd	$= n + 3$
	if not $\div 6$ and odd	$= n + 5$
	if n is anything else	$= n + 4$
$7 \times n$	if $n \div 7$	$= n$
	if n is anything else	$= n + 6$

Figure 11. Condition Statements for $1 \times n$ to $7 \times n$ rectangles

number of rooms entered is 15; second, a progressive pattern occurred across the sets starting with $n + 1$ in $2 \times n$ rectangles, $n + 2$ in $3 \times n$ rectangles, $n + 3$ in $4 \times n$ rectangles, and so on. Therefore, one condition in every set would include n plus one more than what occurred in the previous set. This latter condition was labeled as the "anything else" or "any other" condition into which all remaining numbers fit. Recognizing these consistent patterns across sets changed the role of the specific examples generated. Rather than being the source for pattern-noticing, drawn examples become random spot tests for their conditions. That is, once the condition statements were written, Kylie and Tamera each chose one example with which to test the statements. If the test was successful they moved on.

Although more than two conditions were required in some sets, Kylie and Tamera did not explicitly make distinctions as to which numbers had more than two conditions. Their use of "even" and "odd" as explanatory devices from the body of mathematics may be viewed as bounding the mathematical world they brought forth to a particular conception of the mathematics within it, but their use of odd and even was sufficient for their concerns raised and addressed to this point.

Narrative: Building a Global Image

The final twenty minutes of Kylie and Tamera's investigation may be viewed as a synthesis of the information gathered through investigating the questions raised and the explanations offered in action.

After spending over seventy minutes investigating the Diagonal Intruder, Kylie and Tamera began to wind down. I asked, "So do you have a whole global thing now? If I just gave you a number or something could you figure out all the different conditions?"

Tamera responded, "For the odd ones, yah."

Kylie agreed, "Odd ones are easier—"

"The odd numbers are easy but even numbers have a lot more."

"You can get some of the even stuff, like we had the $n + 5$ and the $n + 4$," Kylie stated, referring to two of the four conditions for rectangles with widths of 6 (see Figure 11).

"Yah," agreed Tamera.

Still thinking about the condition statements for widths of 6, Kylie said, "We had the first three [statements], but I think there is going to be a new one for each time probably," offering a conjecture.

Tamera agreed. "For 8 there would be something new. For 10 there would be something new again. The odd ones seem to be the same kind of thing every time."

"So 9 should be just a 2?" I asked.

"Yah. So it should be—"

Kylie used the information gathered and began to list the two likely condition statements, "Divisible by 9 and plus—"

Tamera finished, "and plus 8. 'Cause it's one less than the number." That is, one less than the constant width of the rectangle (see the first two statements in figure 13).

Realizing this pattern for the first time Kylie said, "Cool! Yah."

"It should be $n + 8$ if it's anything else."

"What's a 9 x 6?" I asked, adding, "I've got a theory."

Using their drawings of a 6 x 9 they determined it was 12. "Oh, but what—? From our theory we would get 14."

I offered the following thought: "I was just thinking that we were dividing it, so whether divisibility has anything to do with it."

They attempted to resolve the contradiction between their 6 x 9 diagram and their condition statement for a 9 x 6.

Kylie asked, "why is our 6 x 9 and 9 x 6 off?"

"I don't know. I think all it is—"

"It is the same. It's just our theory."

Explanations in the Process of Understanding 77

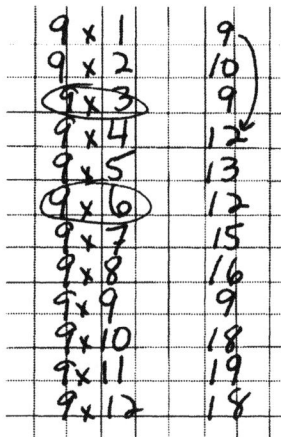

Figure 12. Number pattern for 9 x n rectangles

"—doesn't work" Tamera stated.

"How come our theory doesn't work?" Kylie asked.

"I don't know."

"So where did it go wrong? It went wrong because—"

Tamera offered an explanation, "Maybe 'cause this isn't a prime number anymore. We did 1, 3, 5, 7 and now we are up to something different."

Kylie said, "'Kay, let's do the 9's. 'Kay, 9 x 1 is 9, 9 x 2 should be—"

"Should be the same as 2 x 9, which is what?"

They created a table for rectangles with widths of 9 by using drawings from other sets (see Figure 12).

Their table revealed that 3, 6, "15 and 21 don't work. I betcha anything it does have what Lynn said, 'cause 3 doesn't go into 8. And so it will be where the dots go—will have something to do with it."

Tamera said, "There's another condition for this." They added a third condition: if n is divisible by 3 then the number of rooms entered is $n + 6$ (see Figure 13). Kylie and Tamera numbered the statements according to which condition to test first, second, and then third.

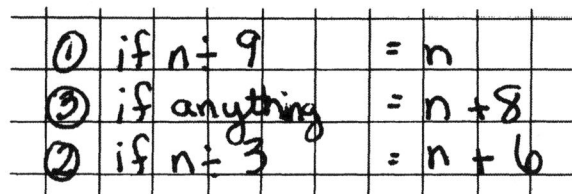

Figure 13. If-then condition statements for 9 x n rectangles

Constant Width	Number of Conditions
2	2
3	2
4	3
5	2
6	4
7	2
8	-?-
9	3

Figure 14. Relationship between rectangle width and the number of conditions

They returned to the earlier question of whether they could predict how many conditions a given set of rectangles would have. In reflecting on this question they created a table of the constant width and the number of condition statements that were needed (see Figure 14).

Kylie said, "'Kay, let me write that down. 2's it was 2; 3 it was 2; 4 is 3."

Tamera added, "5 is 2; 6 was 4; 7 was 2."

"8 we didn't do did we? 9 we found 3. That was the first odd one that went more than 2." Reading down the results for the *even* sets of rectangles Kylie made a prediction, "2, 3, 4. Maybe 8 is 5?"

"How many factors does it have though?" Listing the factors for each number from 2 to 7, Tamera said, "'cause there is 2 and 1; 3 and 1; 1, 2, 4; 5 and 1; 1, 2, 3, 6; 1 and 7; 1, 3, 9. So it should have 4." Tamera's theory was that rectangles with a constant width of 8 should have 4 conditions since 8 has 4 factors.

"Right," Kylie agreed.

"1, 2, 4, and 8." Tamera stated the 4 factors for 8.

"Or it could be going 2, 3, 4, 5." Kylie restated her previous conjecture based on a perceived pattern in their chart (Figure 14). "Now all we have to do is figure it out."

They spent several minutes determining the condition statements for 8 by generating a pattern from previously drawn rectangles. Upon completion, they found that there were four conditions for rectangles with widths of 8:

(1) if n is $\div 8 = n$
(2) if n is $\div 4$ and even $= n + 4$
(3) if n is odd $= n + 7$
(4) if n is even $= n + 6$

Once the statements accounted for all the numbers in their pattern Kylie said, "Yah. So it's 4. So Tamera was right. It's every, it's how many factors they have."

Discussion of Global Image Narrative

Differing Explanations and Counterexamples

This final segment of Kylie and Tamera's investigation reveals, perhaps for the first time, different explanations within the conversation. One conjecture that both Kylie and Tamera agreed to at the beginning of this segment was that odd numbers had two condition statements. I suggested they look at rectangles with widths of 9 and offered 9 x 6 as a *counterexample* to their condition statements (Lakatos 1976). The counterexample was offered as an occasion for Kylie and Tamera to reexamine the assumptions on which they based their previous explanations. When the 9 x 6 rectangle could not be explained using their present conditions, I offered my own tentative explanation or "theory" as I referred to it: "I was just thinking that we were dividing it, so whether divisibility has anything to do with it." This explanation was stated as a plausible rather than as a definitive or convincing statement. The tentativeness of the statement is also reflected in its grammatical incompleteness. However, the explanation was not acknowledged directly until much later when Kylie revisited it and connected it to their previous actions: "I betcha anything it does have what Lynn said. . . . And so it will be where the dots go—will have something to do with it." My explanation did not mention their "dots," but Kylie connected my explanation involving divisibility to their earlier actions. Tamera also offered a conjecture or possible explanation: "Maybe 'cause this isn't a prime number anymore. We did 1, 3, 5, 7 and now we are up to something different." This conjecture was not offered by Tamera to convince Kylie, but was offered to Kylie and to herself to make sense of the experience. Tamera's explanation was also not directly acknowledged by Kylie. Neither the explanation offered by me or by Tamera was initially accepted by Kylie within the conversation. One interpretation for Kylie's lack of acceptance of Tamera's prime/composite number explanation may have been because she did not understand the explanation offered, but since we cannot determine what Kylie did or did not understand, in keeping with the criterion for acceptability in a mathematical conversation, we may conclude that Kylie did not accept the explanation because it did not broaden her understanding of the topic of concern.

Within the final nine minutes of the investigation, Kylie and Tamera attempted to predict the number of condition statements for 8 and offered two distinct explanations. Kylie continued to use an image of number patterns and odd-even numbers to predict the number of conditions for 8 by pointing to the pattern that was evident in the previous even numbers:

$2 \rightarrow 2$
$4 \rightarrow 3$
$6 \rightarrow 4$, hence,
$8 \rightarrow 5$.

Tamera offered an alternative explanation. She noted that the number of factors for both odd and even numbers correlated with the number of condition statements. Eight had the factors 1, 2, 4, and 8, so she predicted there would be 4 conditions. Both explanations were offered to the other person and to themselves as a means to make sense of the concern that had been raised. Neither person attempted to compel the other person to change her mind. Yet, it also does not appear as though either person attempted to understand the other person's explanation. After Tamera offered her explanation, Kylie responded by saying, "Right," but immediately offered her own explanation to which Tamera did not respond. At this moment it appeared as though they brought forth a multiversal reality. They were "accepting" of the other person's explanation, but unlike a conversarial reality, it does not appear as though they attempted to understand the other's explanation and seek a place of coexistence.

Once both explanations were offered they allowed the mathematical situation to validate which explanation would be accepted. Afterwards, Kylie says, "Yah. So it's 4. So Tamera was right. . . . It's how many factors they have." Although Kylie gave verbal acknowledgment that Tamera's explanation was validated, it may have been necessary for Kylie to continue to engage with the phenomenon further for evidence as to whether this explanation was accepted by her by becoming part of her subsequent actions.

The explanations offered in conversation are offered to oneself and to the other. They allow a person to re-present and share his or her understanding and actions in and for that moment. In a mathematical conversation based on mutual acceptance, it is expected that the listener will respond to an explanation offered in some way—not necessarily by agreeing or disagreeing, but by attempting to understand it by asking questions, rephrasing it or acting on it in some way, perhaps by expanding the explanation. In the latter part of Kylie and Tamera's activities, the continuity of their conversation began to disintegrate as they formulated two different explanations which led to two different images of coherence within the mathematical situation. While Kylie and Tamera's relationship with the mathematics remained intact, their interpersonal relationship appeared to loosen. This was not due to the fact that two different explanations were offered, but because there was a lack of response to the explanations that were offered—there was no responsive attempt to understand.[6]

Although Tamera and Kylie's investigation could have continued on much longer, they were satisfied with their efforts at the conclusion of this session. Their satisfaction was not that they had solved a problem, but that they had addressed the concerns raised throughout their activities and they were satisfied with their explanations to that moment. Their conversation ended or disintegrated when they no longer raised additional questions of concern.

Explanations within a Mathematical Conversation

It may be useful to distinguish between two forms of mathematical explanation: *Explanations in action*, and *explanations as re-presentation*. The form of explanation that seemed prevalent throughout Kylie and Tamera's investigation were *explanations in action*. Kylie and Tamera's explanations occurred in action, in the midst of trying to understand a mathematical phenomenon brought forth through their interactions with one another. Explanations in this context are not a method for understanding, but are an event of understanding; an event over which one has only partial control (Johnson 1987). Rather than convincing the other of a particular truth as an endpoint, their explanations in action allowed them to say or to experience a feeling that they could "go on" (Wittgenstein 1953). Tamera explicitly said, "Let's go onto 3's" when they prepared an acceptable mathematical explanation at the beginning of their investigation. An explanation that is accepted as mathematical, plausible, and coherent and which broadens the conversants' understanding of the situation allows them to maintain their relationship with one another and with the mathematical phenomenon brought forth.

Explanations in action are an offering or an invitation to oneself and others. They are posed with one's self and the other fully present, are responsive to the other, and are spoken with the intention that they will be heard and viewed as sensible. The explanation may invite images, models, metaphors, and narratives of the topic of concern. Kylie and Tamera's explanations were not attempts to convince the other, but were offerings which expanded their understanding of the phenomenon. In expanding their understanding they also generated new ideas and enriched ideas previously stated. The goal of explanations stated in action is not to reach a particular endpoint, but to provide a means for understanding the phenomenon for and in that moment which allows the participants to maintain their mathematical relationship and continue their conversation.

Mathematical explanations in conversation and stated in action, appear to bring forth either a multiversal or conversarial reality. The conversants recognize that their explanations are not certain or complete. In both realities, a person's actions and interactions are fully implicated in the explanations one offers. Explanations arise from the participants' experiences, rather than from a world unaffected by their interactions within it, and are validated, as we witnessed with Kylie and Tamera, by referring to their actions and experiences within the setting.

Explanations as re-presentation are potentially another form of explanation that may be evoked when a person is asked to summarize his or her thinking to persons outside of the conversation, perhaps to the teacher as she passes by, or to other students as they share their mathematical solutions and explanations with one another. Explanations as re-presentation may have the potential to bring forth a universal, multiversal, or conversarial reality; however, the emotional predisposition from within which explanations are stated needs to be acknowledged. Explanations as re-presentation offered "into" an argumentative discourse are expected to be convincing arguments of one's actions and are responded to by others through either agreement or disagreement according to some explicit or implicit criteria for acceptance. Proofs are an example of explanations as re-presentation and they are offered to and judged by an audience using criteria of perceived accuracy and completeness. As was mentioned in the previous chapter, stating an explanation with a desire to convince may unintentionally shift the emotional predisposition from one of understanding the other, to one of convincing the other; if this occurs, the reality brought forth may shift from a multiversal or conversarial to a universal perspective. When a universal reality is brought forth only one explanation is acceptable as only one answer can correctly match the "real" world. If an explanation brings forth a universe and convincingly appears complete and certain, the conversation may draw to a close. Needing further examination in classroom settings are the ways in which explanations as re-presentation can be offered to peers and the teacher outside the initial conversation so that participants can continue to bring forth a conversarial reality allowing them to continue to reflect on the situation rather than bringing it to an end.

Explanations in action are tentative, plausible, and good enough for the temporal now; however, mathematics as a discipline has had difficulty acknowledging the role that formative explanations play in learning mathematics and in expanding fields of study within mathematics. Explanations occurring in action, in the midst of understanding, continue to be elusive. While argumentation as a form of discourse appropriately provides a means for preparing and critiquing complete explanations for bounded problems, what we as educators also need to be aware of is those "behind-the-scenes" aspects of mathematical practice. These include explanations occurring in action—in the process of making distinctions on mathematical objects, understanding those objects brought forth, and expanding the boundaries of a mathematical field of play.

Mathematics as an Explanatory Domain

Mathematics may be thought of as one particular *explanatory domain* of the questions we ask and the explanations we offer for these experiences. The body of mathematics as an explanatory domain is defined and determined not by a particular set of content or subject matter, but by the mode in which explanations are posed and accepted as reformulations of lived experiences (Maturana 1988). The logical coherence of an explanation depends on reason, but prior to its acceptance by a community it must coincide with the prejudgments, beliefs, and assumptions of that community. Proponents of various ideologies in science, religion, politics, or philosophy often make their arguments in the name of logic, reason, and rationality; however, arguments between factions may not be conflicts of coherence, but are conflicts related to the emotional and underlying assumptions upon which the argument is based. A consequence of emotionally based arguments is that "we cannot force anyone, through reason, to accept as rationally valid an argument that he or she does not already implicitly accept as valid. To move to accept an argument, one must change his or her emotional basis in listening" (43). Although we can attempt to explain with a desire to convince, the other must already accept either the prejudices upon which the argument is based or he or she must accept the authority of the arguer.

A mathematical explanation that is accepted is not one that corroborates an independent reality, but is accepted by someone, the listener, who listens for satisfactory or adequate expressions of knowing within the explanation in relation to some criterion of acceptance (Maturana 1987). Mathematical knowledge and understanding are attributed to the conversants by observers and by the conversants themselves when their actions are viewed as adequate according to implicit or explicit conditions or criteria of acceptability (Maturana 1991). As our culture arises and evolves through networks of conversation, we come to agree on the criteria for accepting explanations in various explanatory domains. Individuals who use the same criteria for generating, corroborating, and accepting their explanations act within the same speech genre or discourse and listen with similar emotional prejudgments; thus they operate in intersecting explanatory domains.

Tamera, Kylie, and I were operating in intersecting explanatory domains utilizing similar aspects of the body of mathematics when we all accepted the if-then condition statements as an appropriate form of explanation for predicting the number of rooms the Diagonal Intruder entered in a 2 x n hotel. Explanations were accepted when they provided enough of an understanding of the immediate situation to allow the conversation to continue. Although the explanations were accepted by Kylie and Tamera in plausible and necessarily incomplete forms, if the explanations were offered to a broader mathematical community they may not have been acceptable if the criteria used for accepting explanations changed.

Mathematics exists as a network of practices and forms of life arising as a particular domain of explanations in which the community within this practice (e.g., mathematicians, teachers, curriculum developers and students) listens and looks for what they consider are mathematically acceptable reformulations of their lived experiences. A consequence of this set of similar questions, assumptions, and criteria is that accepted explanations subsequently define what mathematical practices are legitimate and what forms of discourse are appropriate. For this reason, if mathematical explanations in action are to be accepted by the mathematics community as mathematical, the community must alter its criteria for acceptance. Until then, "mathematical explanations in action" will not be viewed as mathematical or as an acceptable mathematical practice in and of itself—only as a precursor to "real" mathematics. If such acceptance does occur, explanations in action may open the practice of mathematics to all persons and allow students and teachers to see mathematics as an integral part of their humanness and as one domain through which to find coherence and meaning in their experiences.

Notes

1. An overview of the research project and its participants is available in Appendix A.

2. The Diagonal Intruder also appears in the data for chapter 6. The participants in this chapter brought forth a geometric image of the Diagonal Intruder; hence, very different questions and explanations occurred.

3. Both Kylie's and Tamera's diagrams show their physical counting. Kylie shows her counting by loosely shading in the squares entered by the diagonal. Tamera places noticeable marks on the squares counted (see Figure 10).

4. Of course, the students' histories of interaction with teachers in the past may create an emotional desire to shift explanatory paths from a conversarial to a universal reality when a teacher attempts to enter the conversation. That is, students may believe that a correct or complete answer already exists and that the teacher likely already possesses that answer. In these cases, the teacher's "mere presence is often enough of an interruptive perturbation that the conversation disintegrates into a mere *collection of individuals*" (Sawada 1991, 359).

5. Although their actions did not change, the means and format for explaining changed between the sets defined as "squares" and those defined as "$2 \times n$s." The acceptance of the $2 \times n$ if-then conditions, however, did redefine their domain of actions and acceptable explanations.

6. The following chapter addresses the interpersonal and mathematical interactions in more depth.

Interlude

Improvisational Acts

> I believe that our aesthetic sense, whether in works of art or in lives, has overfocused on the stubborn struggle toward a single goal rather than on the fluid, the protean, the improvisatory. (Bateson 1989, 3)

Improvisation in theater is occasioned by an initial dramatic prompt which may be a word, an image, or a situation. The plot that unfolds is a creation of spontaneous action, interaction, and communication that is unique not only to the audience who watches, but to the actors themselves. Although the improvisation is spontaneous and unpredictable, it is by no means random. Each actor brings into the situation a history of knowing and experience and, as the drama unfolds, the actions they take and the roles they play are reflections of their individual histories in interaction.

The improvisation is simultaneously occasioned and constrained by the initial situation and the actors' knowing and experience, but it is also occasioned and constrained by the environment the actors are in, which includes the play-space: the props available and the actors' awareness of the audience. The audience may be regarded as "a many-headed monster sitting in judgment," or a "looker-in" that needs to be tolerated, or as "a group with whom [they] are sharing an experience" (Spolin 1963, 13). There are even times when the audience members are invited on stage to become part of the action. In all cases the actors are not only trying to make the play believable to the audience, but also—perhaps more importantly—they are trying to make it believable and authentic for themselves. Some of the most appealing improvisations occur when both the actors and the audience have lost themselves in the play.

It would be ridiculous to assume that there is a hidden script that is available to all of us in the audience, but not to the actors themselves. As a single member of the audience or potential audience, it would also be foolish for me to think that the interactions that take place are really insignificant and to think instead that what is important is that it ends according to my prior expectations. If either situation were true I would naturally be compelled to make judgments of good or bad, and right or wrong according to these preset conditions.

But neither of these situations is appropriate to the improvisation. Instead, I watch, sometimes having no idea what the actors are about to do from moment to moment. But just as they bring their histories of experience to the improvisation, I bring my own. I have seen a large number of improvisations in the past, sometimes with identical or similar starting points, and I am also aware of many dramatic techniques that may be used. I may be able to make some inferences about where the play is going and whether it is more or less appropriate given the initial and evolving situation, but I am often surprised. Many improvisations that I thought were heading down some familiar path led me to places I had never been or even considered. At these times I often turned to the people beside me to see if they understood what the actors were doing or where they were going. Sometimes someone might recognize it as a parody of another play, or someone else explains the technique being used, but many times none of us knows, so we move closer to the edge of our seats to see what happens next.

Chapter 6

Relationships Within Mathematical Conversations

Introducing the Participants and the Prompt[1]

Stacey and Ken sat down at the table in a small workroom surrounded by blackboards, books, and computer equipment. The camera focused on them, but the large microphone on their table also picked up the voices of the two researchers present. The researchers started the session by trying to relieve the tension Stacey and Ken might be feeling. "We aren't going to psychoanalyze or anything like that. We are simply interested in how people learn mathematics."

Stacey, a mathematics major in education, had volunteered to participate with her friend Ken, a business student, as her partner. The Arithmagon prompt was placed on the table between them (see Figure 15, adapted from Mason, Burton, and Stacey 1985). One of the researchers handed them a sheet of paper and said, "This is an outline of today's activity."

"You will notice that we have given you only one set of materials because we are hoping that in this case you will talk to each other."

Narrative: Solving a System of Equations

After Stacey and Ken both read the prompt in silence, Ken turned to Stacey, "I guess this is uh—"

"Trial and error?" she suggested.

"No. We'll go—Let's assign variables to these then," pointing to the vertices of the triangle.

"Go for it," Stacey's encouragement also implied that he do the calculations.

> **The Arithmagon**
>
> The numbers on the sides of this triangle are the sums of the numbers at the corners.
>
> Find the secret numbers.
>
>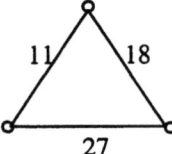
>
> Generalize the problem and its solution.

Figure 15. The Arithmagon prompt

"We'll start with A, B and C," he said, labelling the vertices. "And I guess we'll probably get a system of equations out of that. OK. So that's the sum of A and B" pointing to the side of eleven.

Stacey read off the three equations and Ken recorded them, "A and B is 11. B and C is 18. And A and C is 27."

Ken proceeded to solve the Arithmagon as a system of equations with 3 variables and 3 unknowns. Although Ken did all of the calculations, here and throughout the entire session whenever this method was used, Stacey was involved *physically*, watching and pointing to the Arithmagon and his writing, and *verbally,* checking his work. Within a minute they found an answer, transferred their answers to the Arithmagon and verified their solution (see Figure 16).

"Now we're supposed to play with this?" Stacey asked.

"We have to like generalize it." Ken's conception of generalizing here and throughout the session was to didactically explain the procedure they used to solve it. He said to Stacey, the education student, "This is where your teaching part comes in. You have to try to explain to someone else how to do it."

He asked her a few questions about how to teach solving a system of equations and what grade it was taught in before Stacey posed a question leading them temporarily in a different direction. "What happens if you add the middle numbers together?" referring to the numbers on the sides.

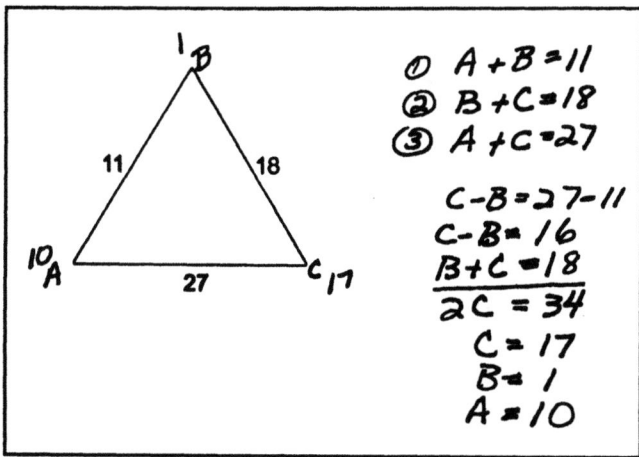

Figure 16. Solution to the Arithmagon

"[O]kay, I just want to try something. If you take 27, 18 and 11" adding them, "56, right?"

"Sure."

"So you add each of those twice, right?" she said pointing to the vertices, "Yeah, you do," answering her own question. "That's not going to help you, either. That's what you end up doing, right?"

"What'd you do?" asked Ken.

Stacey explained the property she noticed within the Arithmagon, "You add A, B, C." That is, 1 + 17 + 10 = 28. "Then you multiply them by two," 28 x 2 = 56. You get this answer" referring to the sum of the sides. In other words, she noticed that the sum of the sides (56) equals twice the sum of the vertices (28).

She checked her calculations again out loud and asked Ken, "Do you know what I mean?"

He responded, "So you add these," pointing to the vertices, "and multiply by 2. So the sum of this is 28 times 2. And it's 56. Good one. What's that mean?"

Stacey laughs, "Nothing."

"Is that true for all of them?"

"Yeah."

"I guess so. It must be. It can't just be fluke." Ken moved their activity in an alternative direction. "OK—I guess. We could also—solve it as a matrix also, I guess."

Even though Ken assumed that they would "do the same steps" as before, they spent several minutes trying to remember how to use matrices to calculate an answer. When the answer calculated was different from their previous one, Ken said, "Did we make a mistake? Or did we just, or is it just that we don't remember the process as well as we thought."

Although they knew the answer was incorrect it prompted a discussion on whether there might be more than one answer. Stacey asked, "What if we put negative numbers in" for the solution?

Ken gave the following explanation, "I think there is only one solution for this, though. There is when you are programming. It's like if you have the same number of variables, if you have the same number of variables as you have equations, then that means that there is only one possible set of answers, and if you have, if you have, more variables than you have equations then there is an infinite number, and if you have less equations than you have variables, then it's ahh, I think, is that also there? One of our solutions?"

"Oh, I never learned that."

"But anyways, I'm pretty sure that if you have three variables, and three equations, that I think there is only one answer. . . . Well, let's get back to generalizing this with the system," Ken suggested.

Still unsure what this meant, Stacey asked, "How would *you* generalize it?"

"Well first, what is the first step we did? We can look at our steps." Ken proceeded to give a lengthy explanation of the procedural steps taken to solve the system.

Shortly after giving his verbal explanation and a few moments of silence he asked the researchers, "How many problems are we supposed to look at today?"

One researcher responded that they would only be given the one prompt today. The other researcher, possibly sensing that Ken may have run out of ideas, said, "If you want us to give you some hints or directions, you can ask us about it. In other words we can think of other things we would have you do."

Discussion of Systems of Equations Narrative

Image of Prompt and Forming a Topic of Conversation

Stacey and Ken are just two of many participants who were given a similar prompt with similar instructions. Encouraging participants to "talk to each other" or "work together" brought about many different forms of interaction. Some pairs chose to collaborate fully, others found ways to divide the labor, while others worked in parallel, working individually but sharing findings throughout their investigation. There are, of course, other pairs who worked entirely separately, perhaps viewing "the goal of solving the problem as more important than the goal of solving it together" (Azmitia 1990, 143). Many of these latter pairs formulated such different images utilizing completely different aspects of the "body" of mathematics that they no longer had a basis from which to converse, even though they often continued to talk out loud to a potential listener.

The increased emphasis on communication and social interaction in mathematics education has encouraged many teachers to create situations where students are expected to "work with a partner" in mathematics. The previous paragraph described a number of working relationships, however, so it is important that we consider what form of interaction is expected and what actually takes place. This chapter focuses on the relationships Stacey and Ken form between themselves and also between themselves and the mathematics brought forth as they engaged in a mathematical conversation. The formation and maintenance of their social relationship in this setting is connected to the relationship they form with the mathematics; simultaneously, the image they form of the prompt and the mathematics brought forth potentially affects their interaction with each other.

Much of the first twenty-five minutes of Stacey and Ken's investigation was spent doing calculations using more or less known mathematical techniques. For the most part, their actions were fairly narrowly bounded to a set of procedures. The mathematics that is completed in this section is not unusual. Many students with a background in systems of equations or linear algebra bring forth an image of linear systems in response to the Arithmagon. However, having a standard algorithm to follow, in this or in any other mathematical situation, often collapses the opportunity for conversation. I have observed many instances where participants who initially solved the Arithmagon as a system of equations formed less collaborative working relationships. For instance, some participants worked in parallel, solving the system independently and then verified their solutions by checking with one another upon completion; or in other instances, one person solved the system of equations while the other person waited or began working on a different issue of concern. Forming a mutual topic of concern is difficult when the participants have an image of the prompt as a problem that is clearly defined with a foreseeable endpoint. Having such an image often leads to a goal of finding an answer as efficiently as possible, rather than a goal of understanding the phenomenon more thoroughly by raising their own questions and concerns. The students interact with these problems as "rituals" brought forth from their experience and expectations in previous classroom communities.

Once an answer has been found to what seems to the participants to be a clearly defined problem, a mathematical conversation is often difficult to sustain. The activity of Stacey and Ken in the previous narrative appears easily segmented into "problems" that are only loosely related. First, they solve the Arithmagon using algebraic techniques; next, Stacey notices a property that the sum of the sides is twice that of the vertices; however, this property is abandoned and they then attempt to find a solution using matrices; finally Ken generalizes the solution by providing the procedural steps used to solve it as a system of equations. With no other pending problems or issues, Ken appears somewhat at a loss for what to do next. Ken and Stacey had been informed that most sessions lasted approximately one hour. Their activities in this segment took only twenty-five minutes. At this point Ken asks the researchers, "How many problems are we supposed to look at today?" This suggests that he views the Arithmagon problem as solved and generalized adequately. Perhaps the expectations arising from his recurrent interactions in past mathematical communities are for his engagement with problems to cease when solutions have been obtained, when a sense of certainty has been reached.

Although this section of Ken and Stacey's investigation was completed doing calculations to a foreseeable endpoint, the potential for conversation exists in their commitment to working collaboratively with one another. They demonstrate their intentions to engage in this activity together. These intentions are observable in their gestures and words, in their actions and interactions.

Intimate Spaces
As Stacey and Ken engage in this investigation, their histories of interaction become braided together in the conversation; they remain distinct as individuals but they become coupled in the interaction. What emerges is an improvisation or conversation that is choreographed in the moment through the "consensual coordination of action" (Maturana and Varela 1987). Such action in conversation is not merely coordinated, but it has a rhythm, a flow (Taylor 1991). Although defined as "consensual," the interaction is less a consensus or agreement on a course of action and more an agreement to maintain their relationship and the rhythm of their interaction (Tomm 1989). In many ways the consensual coordination of ac-

Figure 17. Stacey and Ken

tion can be described as *co-sensual* action; actions sensually guided by our perceptions and interactions with others and otherness. There is an "intimate reciprocity" in our co(n)sensual actions (Abram, 1996, 268).

The still photo above, captures Ken and Stacey's bodies and the "togetherness" of their working and social relationship (see Figure 17). They work beside one another referencing the same drawings and symbols on paper. As they work side by side their intimacy with one another is revealed. It is an intimacy that is carried forth from their social life to this mathematical setting. Their commitment to one another stems from a history of caring that they bring to this mathematical setting and it is a commitment that cannot be separated from their activity.

Perhaps doing mathematics with another, particularly in conversation, requires some level of intimacy. Individuals must be willing to take risks, to trust the other, to listen to and respect ideas generated. A mathematical conversation requires that individuals work closely, within the same space, rather than across the table with a barrier between their bodies. Being too far removed from this interactive and interpersonal space, and the communicative tools of mathematics perhaps precludes the possibility for a collaborative relationship in mathematics. The mathematical space is necessarily close, intimate. Although their bodies create distinctions between them, Stacey and Ken's bodies also place them in relation with each other and with their surroundings. As Abram (1996) reminds us,

> The boundaries of a living body are open and indeterminate; more like membranes than barriers. . . . Certainly, it has its finite character and style, its unique textures and temperaments that distinguish it from other bodies; yet these mortal limits in no way close me off from the things around me or render my relations to them wholly predictable and determinate. On the contrary, my finite bodily presence alone is what enables me to freely engage the things around me. . . . Far from restricting my access to things and to the world, the body is my very means of entering into relation with all things. (46–47)

Both Stacey and Ken have entered into relationship with each other and with the mathematical task they framed with an intention of caring and of maintaining their relationship with each other. Their experience is brought forth as an act of *love*. Love "lets us *see* the other person and open up for him room for existence beside us. . .without love, without acceptance of others living beside us, there is no social process and, therefore, no humanness" (Maturana and Varela 1987, 246). Love, defined here as the mutual acceptance of the other, is a necessary element in conversation. It is a stabilizing factor in all social communities (Maturana and Varela 1980). It is through an emotion of love that a conversarial reality is brought forth. The histories of individuals intertwine as they lay a path together. It is through love and within this intimate space that one's horizon for understanding can be expanded.

The intimate space required for a mathematical conversation potentially limits the number of people who can fully engage in a mathematical conversation at one time. Although my observations have been primarily with student pairs, groups of three have occasionally occurred. In such instances, one person, the one sitting on the end or on the other side of the table frequently could not participate fully simply because he or she could not see the mathematical references put on paper. Communication of mathematical ideas is in large measure nonverbal in terms of its reference to symbols, diagrams, and drawings on the page, as well as the gestures which include pointing to, as well as talking about. Talking about mathematics without physical gestures and symbolic references would be extremely difficult. Perhaps "talking about mathematics is like dancing about architecture" (H. Kass, personal communication, spring 1995). Creating a space for mathematical conversation requires attention to the physical space, as well as opening the linguistic space for student voices.

Cultural/Gendered Role-Playing
Many of my colleagues who have viewed the interaction between Ken and Stacey have commented on the gendered roles that each person plays. Stacey's ideas, particularly at the beginning, appear rejected by

Ken. Stacey provides a potential direction when she suggests, "Trial and error?" Ken appears to reject this suggestion and offers an alternative course of action when he says, "No. . .Let's assign variables." As a colleague, Dr. Brent Davis (1995a), remarked in response, if we ask "why does she subordinate her thinking to his" we have to consider that they are "placed in the current cultural context (and, in fact, in the pervasive mathematics context). I'd actually have been surprised if the reverse had occurred." Their actions and interactions cannot be separated from their personal histories with one another and the cultural context that they bring to this situation.

While Stacey's ideas have been occasionally placed on hold, it is important to note that Stacey is not sitting back waiting her turn. She actively responds to and acts on Ken's suggestion to solve the Arithmagon as a system of equations. Although Ken keeps the symbolic records for solving the system, Stacey is not "following along," but is actively participating beside him. Stacey often states the symbolic steps before Ken records them and she occasionally corrects his errors. This indicates that Stacey could have completed the calculation on her own. However, Stacey's willingness to interact with Ken as he completes the calculations indicates support for this path of action and for them to continue interacting. They *work together*, not out of necessity or efficiency, but out of a commitment to one another to maintain their mathematical and social relationship in this setting. Stacey's act of listening and responding to Ken's suggestion for "assigning variables" enters them into an inviting conversational space. It is often the listener who creates the space for conversation. Without listening and responding Stacey may have proceeded down a different path using trial and error while Ken proceeded with a system of equations. If such actions had occurred, the space for conversation would have been lost, at least for that moment.

Re-Pairing the Interaction

Stacey and Ken have made room for the other, but this space must be continually re-paired when misunderstandings, confusion, or conflicts arise. By using the word *re-pair*, I am not suggesting something has been broken and that there are tools available to "fix" that which has gone wrong, but rather that a purposeful effort is made to reenter into conversation with one another—to pair again. It is a rejoining of personhoods through language and interaction. Such re-pairing is necessary as "the first condition of the art of conversation is ensuring that the other person is with us" (Gadamer 1989, 367). There are moments in Stacey and Ken's conversation where one person begins a new exploration independently. For example, at one point Stacey says, "I just want to try something," and she begins down a path alone. She looks for a relationship between the sum of the sides and the vertices of the Arithmagon. At one point she says, "So you add each of those twice, right?" Although she speaks out loud and adds the question, "right?" to her comment, it is questionable that she was addressing Ken. However, as the listener, Ken responds to her question and re-pairs their interaction by asking, "What'd you do?" In conversation, we often recognize that we do not understand the other person's ideas, but we wait for something said later to make sense of what was said earlier. Until the idea becomes vital to the conversation we carry on. Stacey's question, "Right?", which may have been asked of herself, perhaps was a signal to Ken that the idea had become vital to the topic of conversation and he could no longer wait for the future to make sense of her past activities.

Ken's response, "What'd you do?" encourages Stacey to explain her course of action so that he could share it with her. Stacey proceeds to explain the relationship between the sides and vertices more thoroughly, both to herself and Ken. She completes her explanation by addressing Ken directly with the question, "Do you know what I mean?" He responds with his interpretation, "So you add this and multiply by 2 so, like, the sum of this is 28 times 2. And it's 56. Good one." From an observer's perspective we may assume that their understanding is "taken-as-shared" (Cobb, Yackel, and Wood 1992). This claim is not meant to conclude that their mathematical understanding is identical, but that they share a similar image that enables them to continue on. "Speaker and listener understand each other not because they have the same knowledge about something, and not because they have established a likeness of mind, but because they know 'how to go on' with each other" (Carse 1986, 131). Reestablishing their interactivity through questions addressed to the other, such as, "What'd you do?" and "Do you know what I mean?" act as occasions to re-pair and to reenter into conversation with one another. Re-pairs to their interaction, evident throughout the session, point to their commitment to maintain their conversation and relationship with one another.

Narrative: Exploring Nested Arithmagons

Twenty-five minutes had elapsed since they began their investigation. Ken turned back to Stacey after having been told that they would only be looking at one problem. Stacey had been looking intently at the Arithmagon

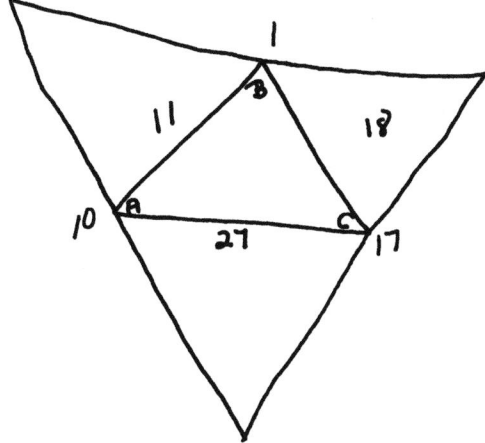

Figure 18. Nested Arithmagon

while the exchange between Ken and the researchers occurred. She redrew the original 18-11-27 Arithmagon on a new sheet of paper and said, "This is what we have: 18, 11, and 27. And we're given three other numbers, right? What can we do with three other numbers? We can, extend lines—" Stacey proceeded to draw an inverted triangle outside of the given one (see Figure 18). From this point on, their investigation became an exploration of nested Arithmagons and the properties arising from these objects.

"Wha'cha doing? Making another big triangle?" Ken asked.

"Yeah. I don't know what I'm doing yet."

Stacey finished labelling the two nested Arithmagons and said, "You keep going."

"Where you going?"

"You keep going. We could find numbers for this," pointing to the outer triangle.

"For that?"

"Yeah."

"Okay. So you want to solve that then?"

"Umhmm," she agreed but hesitated. She continued to look at the Arithmagon and then made a prediction, "It will go right to zero."

"Are you saying—"

"This—I don't, I don't know."

"Are you saying the numbers would keep getting smaller?"

"Yeah, these would have to be—"

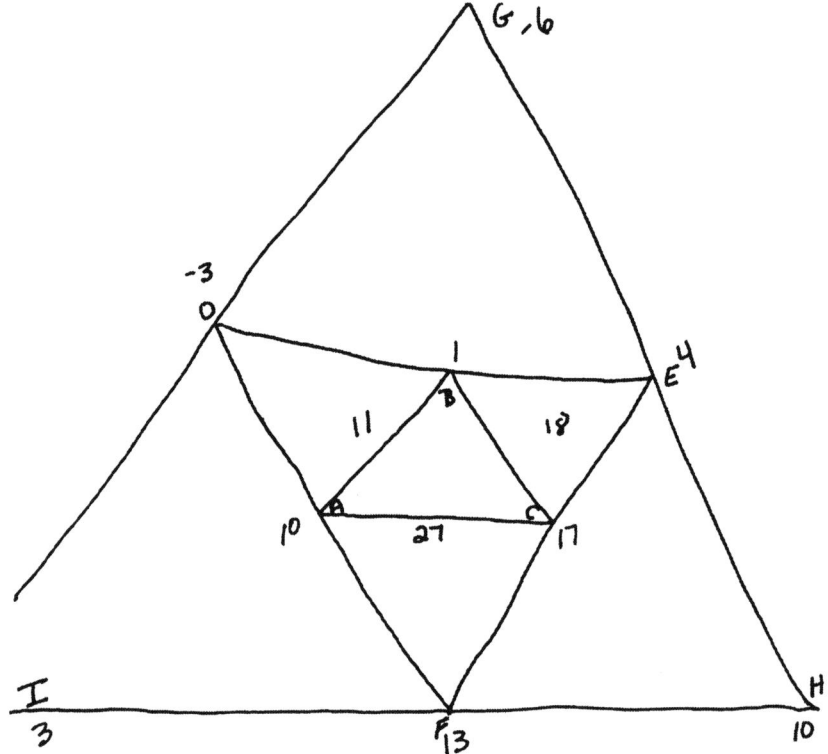

Figure 19. Three Nested Arithmagons

They both agreed that the numbers would get smaller on the outer triangles, then Ken asked again, "You want to try to solve it then?"

"Yeah, sure. Away you go," and pushed the paper to Ken who proceeded to label and solve this new system of equations while Stacey participated verbally in the calculations.

When they found a solution Stacey suggested that they "keep going" by adding a third Arithmagon that did not quite fit on the page (see Figure 19).

As Stacey placed their solution of −3, 4 and 13 on the second Arithmagon (see Figure 19), she said, "Do you know what?"

"What?"

"[O]kay. You want a prediction?"

"Okay. Sure."

"Well, I don't know about this. This is just one little step. This is decreased by 14," pointing to the relationship between one side of the original Arithmagon (18) and the vertex of the outer Arithmagon (4). "This is decreased by 14," from 11 to −3, "and this is decreased by 14" from 27 to 13.

"What's that mean?"

Stacey continued her observation, "from here to here."

"I see that. That's pretty neat."

"Does that mean this," point H (see Figure 19) "will decrease by 14?. . .Is that a prediction that H is 3? Go for it."

Ken proceeded to solve it using a system of equations. When he finished solving for the different variables he said, "H equals 10."

"That isn't at all what I predicted." But Stacey suggested they "keep going."

When they placed the solution on the drawing they noticed that the relationship between the sides of the second Arithmagon and the vertices of the third was a difference of 7 (see Figure 19). Ken said, "So that's 7, that's 7, and aaah!"

"Ooooh!"

"How do you like that? And that's 7. So are you willing to bet" the next one out is, "3.5?"

They solve for a fourth Arithmagon using a system of equations and Stacey says, "That works!"

"And didn't I say 3.5 that one?"

"No you didn't say nothing!" Stacey teased.

"Yes I did! I said it would be minus 3.5," Ken laughed.

"How'd you do that?"

"Well because the first one was minus 14, the next one was minus 7 and this one I thought would be minus 3.5. So let's try, subtract 1.75. That's half of 3.5, right?"

They verified that this relationship held for five nested Arithmagon and it also worked going inwards by multiplying by 2.

"That was pretty cool. So after that now we don't need a system of equations," Ken said.

Discussion of Nested Triangles Narrative

Mathematics as an Imaginary Landscape

The nature of Stacey and Ken's mathematical practice changes in the nested Arithmagon narrative. Questions arise naturally from their mathematical interactions with the object brought forth. There is a flow to their questioning and the explanations of their experiences that was not present in the first part of their investigation when the Arithmagon appeared as a bounded problem to be solved. In the linear equations segment of the narrative, Stacey and Ken appeared to bring forth a universal reality in which they attempted to match their procedural method and solution to a method and solution that existed prior to their actions. In the second segment of the narrative a conversarial reality appears to be brought forth. They no longer look to external sources to justify their explanations, as Ken appeared to do when he appealed to the content of his linear programming class to aid his argument that if there are three equations and three unknowns there will only be one solution. In the nested Arithmagon narrative the source for validating their explanations is their experiences within their conversation. They see themselves as the creators of the mathematics brought forth and do not attempt to recall from past communities a framework from within which to act. Meaning and coherence are shaped as they engage in a conversation between themselves and the mathematics before them.

A conversarial reality is brought forth in their actions within emotions of mathematical curiosity, mutual acceptance, and moral and ethical responsibility to one another. Stacey and Ken's activities in this second portion of the narrative cannot be viewed as "progressing" towards a goal effectively or efficiently as might be expected within a universal reality where certainty is perceived as possible. Engaging in mathematical conversation within the emotional grips of curiosity and acceptance brings forth a conversarial reality which suddenly expands their possibilities for action and understanding.

A "what would happen if" curiosity seems to drive their conversation. This curiosity arose from a question Stacey posed: "What can we do with three other numbers? We can extend lines—." Stacey and Ken's methodology here and throughout the remainder of their investigation became more experimental, involving a cycle of predictions and tests using specific examples; they *become* mathematicians. They bring forth a mathematical world in interaction with the phenomenon. "These constructions, these imagined worlds, then impose *their* order on us" (Davis and Hersh 1981, 406). Yet, the Arithmagon, like the "body" of mathematics, is not static; it is not brought forth as a whole. It is incomplete and exists as a topic of conversation within an imaginary network of potential space, an imaginary landscape.[2] The landscape is recognizable, but indefinite— an ongoing living phenomenon. On the one hand, Stacey and Ken anticipate what is appropriate within this landscape and "act responsively 'into' [the] situation" according to what the Arithmagon openly allows rather than prescribes (Shotter 1995, 95; Varela 1987). On the other hand, the mathematical landscape responds to their actions. Their construction of nested Arithmagons suddenly changes the mathematical landscape into which they act. Stacey and Ken are an integral part of the landscape.

The landscape has a horizon which gives the appearance of a boundary, which may be narrow or expansive depending on one's perception of it and the mathematical phenomenon; yet with each act, with each step, the horizon changes, opening up new possibilities at each moment (Gadamer 1989). In the first part of their investigation, Stacey and Ken's "horizon" was somewhat narrow. They had difficulty seeing beyond the narrow goal of finding a solution using a set of procedures. Yet, the act of "extending lines" broadened the horizon from within which they acted. The landscape changed and the horizon moved with their growing understanding of the mathematical phenomenon.

It is important to note the role that mathematical procedures had in this experimental portion of their investigation. Although the algorithm for solving a system of equations provided a means for generating solutions for the Arithmagons, it was not an issue driving their conversation. Rather than performed or calculated as a goal in itself, the procedure used for solving the system of linear equations became the means through which Stacey and Ken could engage in and continue their exploration. This view provides an alternative possibility for the role that the cultural tools of mathematical practice play in the mathematics classroom:

> When we see technique or skill as a "something" to be attained, we again fall into the dichotomy between "practice" and "perfect," which leads us into any number of vicious circles. If we improvise with an instrument, tool, or idea that we know well, we have the solid technique for expressing ourselves. But the technique can get too solid—we can become so used to knowing how it should be done that we become distanced from the freshness of today's situation. (Nachmanovitch 1990, 67)

The procedure for solving a system of equations initially restricted Stacey and Ken's investigation to a predefined goal, and once they reached that goal there was hesitation about what to do next. When certainty is reached one does not think to reflect or ask further questions (Maturana 1998). Yet, the procedure later became a tool for "expressing themselves." The technique furthered their investigation of nested Arithmagons. As a whole, it was their own curiosity, not a mathematical procedure, that launched them into "the freshness of today's situation." This suggests that the need and use for mathematical conventions are not eliminated in conversation, but they serve a different purpose: "To do anything artistically you have to acquire technique, but you create *through* your technique and not *with* it" (Nachmanovitch 1990, 21). Rather than becoming a goal in themselves or for some future purpose, as they often appear in school mathematics, mathematical procedures and conventions become a potential tool for exploration. They provide a broader means from which to gain an understanding of their experiences.

The Path of Exploration
The direction that Stacey and Ken's investigation leads is unknown both to Stacey and Ken and to the researchers observing their actions. Other than an awareness of where they've been and where they presently are, and a purpose leading them into an anticipated future, Stacey and Ken do not have a final destination in mind. Although there is no agreed upon structure to their exploration, once they begin, there *is* a structure to their actions. The mathematical landscape is bounded by what the mathematical phenomenon will allow and also by the individual histories of experience that Ken and Stacey bring to the conversation. The nested Arithmagon is brought forth through the interactions between persons and the environment.

Although the nested Arithmagon may be viewed as Stacey and Ken's spontaneous and random creation, such a view would belie the experience and history that both Stacey and Ken bring to this situation. As this mathematical conversation unfolded—as they laid down the path of their exploration—who they are, what they believe, and what mathematical skills and interests they have brought are recognizable in what they do, what they say, and in the roles that they play. Although we may feel the need to dissect and identify aspects of the conversation as "belonging" to either Stacey or Ken, it is perhaps inappropriate to say that they are acting autonomously. Their conversation, as is any mathematical conversation, is not a patchwork of individual products, but a fusion of skills, ideas, interests, and experiences occasioned and brought forth in the moment.

> The work comes from neither one artist nor the other, even though our own idiosyncrasies and styles, the symptoms of our original natures, still exert their natural pull. Nor does the work come from a compromise or halfway point (averages are always boring!), but from a third place that isn't necessarily like what either one of us would do individually. What comes is a revelation to both of us. (Nachmanovitch 1990, 94)

We notice the idiosyncrasies in their styles. Ken appears to be rather procedural in his method, while Stacey appears more playful. For example, in the systems of equations narrative Stacey asks as one point, "Now we're supposed to play with this?" Yet, the mathematics that is brought forth is unique to Stacey-and-Ken together. The path laid would have been quite different if walked alone; similarly, it would have been different if Stacey or Ken were in conversation with a different person.

Playful Acts
Stacey proposes early on that their investigation is an opportunity to play. After they find a solution to the Arithmagon problem as they originally defined it, both Stacey and Ken are drawn into their exploration through playful acts—both with the mathematical phenomenon and with each other. "Play is the free spirit of exploration, doing and being for its own pure joy" (Nachmanovitch 1990, 43). Engaging in play with a mathematical phenomenon, as opposed to solving a problem, changes the purposes of the people who play.

Azmitia (1990) contrasts the actions of young children engaged in play compared to them solving problems collaboratively. When we watch children play they seem quite skilled at imagining new scenarios, engaging in role playing, resolving conflicting perspectives, and flexibly adjusting their aims and interaction to incorporate new ideas. However, young children often have difficulty coordinating their interaction during collaborative problem-solving tasks.

> These difficulties may be due, at least in part, to the fact that problem solving requires that peers focus their actions and attention toward a specific, single goal, that is, the solution of the problem. Because play allows flexible alteration of goals, children may avoid breakdowns by redefining their aims whenever they encounter an obstacle. (138)

While the purpose of problem solving is often perceived or described as finding a preexisting solution, a solution which often brings the problem-solving activity to an end, the purpose of play is to engage in activity which allows the play to continue. What is intended in play "is to-and-fro movement that is not tied to any goal that would bring it to an end" (Gadamer 1989, 103).

Although the purposes of problem solving and play may be perceived as different activities, they are not distinct or incompatible activities. In fact, when we observe Stacey and Ken's interactions in this portion of their investigation we could claim that they are trying to solve the "problem" of identifying a relationship between Arithmagons that are nested within one another; however, determining this mathematical relationship may be viewed as a purpose "occurring in the now." However, this purpose is not a final destination or an external goal to be obtained; the purpose in play is always changing. The purpose becomes a frame for action, just as play is a frame for action. Play *is* the purpose. "Anybody who has tried to stop some children playing knows how it feels when his efforts simply get included in the shape of the game" (Bateson 1979, 149). As new ideas, information, and questions are formed, they are simply included into the shape of their purpose, into the shape of their play. Stacey and Ken's play is continuous. Once the mathematical relationship is determined their conversation does not end; rather, the relationship becomes a means for them to continue to play, it allows them to *continue on*.

Rather than filling time or passing the time with mathematically related work, Stacey and Ken's actions within this segment are not bounded by time, but embody possibilities for action (Carse 1986). Their activity is endlessly open. Engaging in play for the purpose of continued play allows the conversation to be sustained and the relationships between the participants and the mathematics to be maintained. Questions continue to arise from and through Stacey and Ken's interactions with each other and the Arithmagon. As long as the mathematical conversation continues there is an opportunity for the participants to broaden their horizon of understanding of the mathematical landscape brought forth.

Narrative: Deriving a Formula

After Stacey and Ken identify a mathematically consistent relationship between the nested Arithmagons, Stacey asks, "So can you derive a formula?"

"A formula?"

As they begin to think about what this formula might entail, Ken poses a concern, "This is something I'd like to understand is that why is 14 such a magic number to start with? Why 14 and not, why not like a different number?"

Laughing, Stacey says, "That is such a good question!" They look at the numbers used for the sides of the original Arithmagon, 11, 18 and 27.

At this point Ken returns to the calculation Stacey did in the first segment of their activity. He says, "So you said that was 56," referring to the sum of the sides. "Does that help you out? Divided by this—"

"Yeah, divided by—"

"Divided by 14. What's that?"

"Divided by 4."

Relationships Within Mathematical Conversations

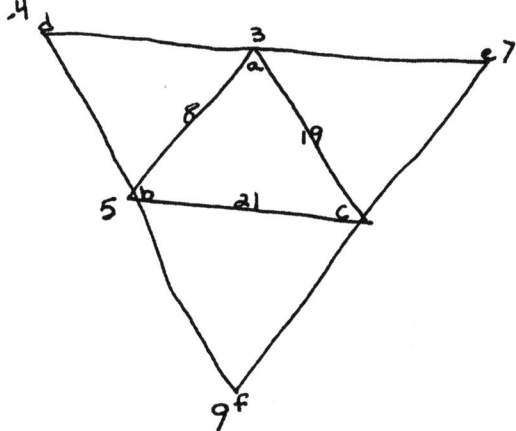

Figure 20. Two nested Arithmagons

"4? Is that 4?" Ken asks.

"What's up with 4? That's great, eh?"

"14, 4, 56. What's the relationship between these numbers?" They both laugh, "Well, obviously it's like that," writing down 14 x 4 = 56, "but why 14 and why 4?"

They continue to look at the relationship of numbers in the nested Arithmagon (see Figure 19). Stacey suggests a course of action, "Do you know what we could try?"

"What?"

"We could try just using other numbers."

"You want to set up a whole new thing?"

"Yeah."

"Okay do it; randomly pick three. Bet you one quarter's pretty important in this one too."

They created a new Arithmagon with sides of 21 19 and 7, using favorite numbers and birthdays. Once they summed the sides they adjusted the side of 7 to 8 which made it more easily divisible by 4. They validated their solution method for solving the nested Arithmagons, by first summing the sides to a total of 48, dividing by 4 which is 12, and then subtracting 12 from each side to find the vertices of the outer Arithmagon to be 9, 7, −4 (see Figure 20).

Once solved Ken says, "Well, we figured out that's how it goes, eh?"

"Yeah, it's quite constant."

Ken restates the procedure and then Stacey asks, "But why did we divide by 4?"

"Because 4 is the—. We figured out 4 is the magic thing here."

"Yeah, but why?"

Although they've figured out a method, Ken says, "But we now have to figure out why 4? So we figured out the system, now we just got to understand *why* the system?"

Stacey offered a geometric explanation relating to triangles. But neither she nor Ken were convinced. They continued to try to understand why 4 was a "magic number" when one of the researchers said, "I think you have been playing for about an hour now," and then asked them to summarize what they did.

Afterwards Stacey commented on the whole investigation. She said, "Actually it was really interesting. I like figuring it out, and adding to it, and going WOW!"

The researchers discussed some of the alternative explorations that other participants had engaged in, such as using square or pentagonal Arithmagons, but one researcher commented, "but I don't think I have seen this before."

Still having several unresolved questions that the researchers could not answer offhand, Stacey smiled and said, "So we'll just take this downstairs now." When Stacey and Ken returned for another session several days later they told the researchers they had gone to the library and continued their exploration with the Arithmagon for several more hours.

Discussion of Deriving Formula Narrative

To be prepared against surprise is to be *trained*. To be prepared for surprise is to be *educated*. (Carse 1986, 23)

Conversation as a Recursive System

Even after an hour of exploration, Stacey and Ken were motivated to continue their conversation in another setting. Much of their activity in the latter half of their investigation was, to use Stacey's words, to "keep going," to maintain their relationship with one another and the mathematics of the situation. Within the second part of the narrative we observed Ken and Stacey attempting to establish a relationship between nested Arithmagons. Once that relationship was validated through the use of examples, a new question was posed and addressed within the final segment of the narrative, stated as, "So can you derive a formula?" which led to yet another question, "Why did we divide by 4?" These questions arising from their own actions placed their activity on the cusp of mathematical formalization. The questions allow their play to undergo another cycle of recursion. There was no predefined goal that they established in their interaction that would bring their exploration to an end. "Rather, [their play] renews itself in constant repetition" (Gadamer 1989, 103).

Mathematical conversations exist in a perpetual state of play. Play is a recursive structure in which the participants' actions fold back on themselves, giving rise to new experiences and new questions. Their play is not random. It has structure, rhythm, and pattern driven by purposes occurring in the now and questions arising from their actions and interactions. The potential landscape occasioned by the initial Arithmagon is vast. Other than finding the perhaps trivial answer to the original Arithmagon, there are no specific problems within it that it instructs the students to address. Thus, we cannot say that Stacey and Ken's understanding of the Arithmagon is complete or optimal, for play is constantly open to new possibilities. Although the prompt and the initial directions given provided a context in which Stacey and Ken can act, their actions and interactions are unpredictable to both the observers and to the participants themselves. Yet, even though they are unpredictable, they are recognizably bounded by the experiences of the participants and bounded by the original Arithmagon itself; that is, the histories of Stacey and Ken are recognizable in the artifacts of their play and the essence of the Arithmagon is also conserved.

The questions and concerns addressed in Ken and Stacey's conversation are not issues in which they eventually share the same understanding, but perhaps they do share similar emotional attachments for the concerns they choose to address. Stacey and Ken's connection to the investigation is based on a continuous sense of curiosity, surprise, and uncertainty. An emotional desire to understand and an interest in knowing what will happen next is heard and seen throughout Ken and Stacey's interactions:

"What can we do with three other numbers?"
"You want a prediction?"
"That isn't at all what I predicted. Keep going."
"So do you want to bet 3.5?"
"Why 14 and why 4?"
"Now we just got to understand why the system [works]."

As their understanding of these questions and conjectures changes, we see and hear their emotional responses in their bodily gestures and in their tone of voice (e.g., "oooh" and "aahhh"). Their actions and their willingness to maintain their conversation were emotionally based. Maturana (1991) suggests that scientists, regardless of claims to emotional independence, proceed on emotion. "The fundamental emotion that specifies the domain of actions in which science takes place as a human activity is curiosity under the form of the desire or passion for explaining" (30). We ask questions that we desire to ask and understand. At the conclusion of the session, Stacey commented on her emotional connection to their investigation, "Actually it was really interesting. I like figuring it out, and adding to it, and going WOW!" Maintaining a conversation with another person requires that a topic of shared significance exists between the participants; a topic which they both desire to understand. This topic of mutual concern allows them to become "bound to one another in a new community" (Gadamer 1989, 379).

Stacey and Ken's exploration suggests a vision of community necessary for inviting mathematical conversations into the classroom. The community that is formed must have the potential to create and maintain relationships based on mutual respect and acceptance between the members of the community, and it must also be capable of sustaining relationships based on curiosity between its members and the mathematics brought forth. Forming and maintaining relationships requires that the community members not only feel a sense of belonging within the community but also feel responsible and accountable to its members and for the mathematics brought forth.

> It is not enough for me merely to have a "place" within [the community]; I must also myself be able to play an unrestrained part in constituting and sustaining it as my own kind of 'social reality,' as not 'their' reality, but as of me and my kind, as 'our' reality. (Shotter 1993, 15)

Inviting mathematical conversations into the classroom requires an acknowledgement of the emotional aspects of interpersonal and mathematical relationships. The members of the community need not only tolerate but must invite ambiguity and uncertainty into their playspace. Rather than being tied to a specific goal, a perceived utility or a final destination, the importance of play for the purpose of continuing to play becomes a key feature of mathematical conversations, but more importantly, it is a feature of the ongoing practice of mathematics.

Notes

1. See Appendix A for a note concerning the research data and interpretations for chapter 6.
2. The concept of imaginary is used by Shotter (1993) to indicate that which "is not yet wholly 'real,' but yet is not wholly 'fictitious' either" (p. 79).

Chapter 7

Lingering in a Mathematical Space

Introducing the Participants and the Prompt

Calvin, who was twelve years of age, and his mother Jolene participated in an extracurricular mathematics program for children ages 8–14 and their parents.[1] Over the course of a ten-week term, Calvin, along with several other children and their parents, came together one evening per week at a classroom in a local high school. For one to one-and-a-half hours each week parent and child engaged in mathematics together. On this occasion "The Diagonal Intruder" was presented (Stevenson 1992, 53).[2] I sat down with Calvin and Jolene as I had done on a number of previous occasions. Professor Elaine Simmt, the instructor of the class and a co-researcher, provided each group with graph paper and drew the Diagonal Intruder diagram on the board (see Figure 21). I explained the day's prompt to Calvin and Jolene: "An intruder keeps breaking into these rectangular shaped hotels. He enters at one corner and exits out the opposite side. The hotel owners are worried and want to know how many rooms the intruder will invade in their hotels."

Narrative: Drawing a Straight Line

Calvin and Jolene sat for a minute in silence looking at the blank graph paper on their desk. Finally Jolene looked up and asked, "What should we do?"

I responded, "Just try different rectangles. Start with some simple ones like squares."

Calvin took the graph paper and drew a 2 x 2 square in the top left corner of the sheet of paper (see Figure 22a). "So then like this?" Calvin asked and traced his finger diagonally across the square.

114   Lingering in a Mathematical Space

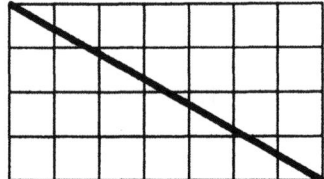

Figure 21. Diagonal Intruder

Jolene responded, "Uh hmm. I think so."
I nodded my head.
Calvin drew the diagonal through the 2 x 2 square.
Jolene asked Calvin, "So how many rooms does he enter?"
"1, 2, 3, 4," Calvin counted the squares.
"4? I thought 2 'cause he just goes through the wall here. I guess it depends on how big he is."
"So 2?"
Jolene turned to me and asked, "Is it 2 or 4?"
"It's your decision. You can choose whichever way you want as long as you're consistent."
Calvin said, "Okay, so 2?"
"Yah, 2." On another sheet Jolene recorded 2 x 2 = 2. "So the Intruder goes into 2 rooms."
Without speaking, Calvin drew a 2 x 3 to the right of the square and drew a diagonal through it (see Figure 22b). "1, 2, 3, 4. 4."
"Uh huh." Jolene recorded 2 x 3 = 4.
Moving to the right again on his graph paper Calvin drew a 2 x 4 rectangle with a diagonal (see Figure 22c). "1, 2, 3, 4, 5."
"5." Jolene then looked at his drawing more closely. "Is it 5? It looks like 5, or does he go through the wall again there," pointing to the center.

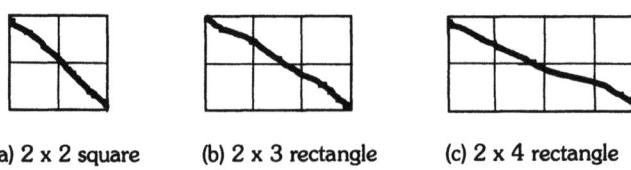

(a) 2 x 2 square   (b) 2 x 3 rectangle   (c) 2 x 4 rectangle

Figure 22. Reproductions of Calvin's drawings

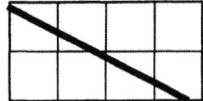

Figure 23. Calvin's 2 x 4 Rectangle

"He's underneath it."

"Is that just the line though?"

Calvin shrugged his shoulders.

Jolene looked around and pulled a dictionary off the shelf beside her. "Here. Try using this to draw your line."

Calvin erased the freehand diagonal line and then tried to line up an edge of the dictionary along opposite vertices of the rectangle. He drew a line. When he pulled the dictionary away he said, "It's not lined up right" (see Figure 23).

"That book's too big. Is there a ruler around?"

"You could fold a piece of paper over a bunch of times." I tore off a piece of graph paper and folded it to make a straight edge.

Calvin erased the line drawn with the dictionary and then lined up the paper straightedge. As he drew, the paper wasn't sturdy enough. His line curved as he pressed his pencil heavily against the paper straightedge. "That didn't work," he said.

"Elaine, do you have a ruler?" I asked the instructor.

"Oh yah. Just a second." She quickly returned with a ruler.

Calvin erased the line drawn by the paper straightedge. He put the ruler in place and held the top of it with his left hand. As he drew the diagonal, again pressing heavily against the ruler, it shifted out of place part way across the rectangle. "Ahhh. I can't do this!" Frustrated, he reached for his eraser yet again.

Jolene and Calvin sat looking at the rectangle for several seconds. Jolene said, "It has to go through the center doesn't it? Because we are kind of cutting it in half so it has to go through the half. See Calvin?"

"So he goes through the wall there."

"Yah, he has to 'cause that's the center," pointing to the middle of the rectangle.

"Okay." Calvin placed a dot on the center and drew the diagonal line freehand through the dot. "So, 1, 2, 3, 4. 4?"

"Yah, I think so." Jolene recorded 2 x 4 = 4.

Discussion of Drawing a Straight Line Narrative

Embodied Actions

For every mathematical act that Jolene and Calvin performed, aspects of the networked "body" of mathematics were brought forth or pulled into action. Jolene and Calvin brought to action mathematical networks related to number through acts of counting and record keeping and Euclidean geometry through the use of geometric properties of the diagonal line and rectangle. The potential for function relationships was also evident in the records Jolene was beginning to keep. Interconnected networks within the body of mathematics were pulled into action and lived through Jolene and Calvin in the present; yet, these networks connect the present action to a historical past. For example, we can imagine Calvin's present act of counting connected to his early acts of counting in interaction with perhaps his parents and his kindergarten teacher. We can also imagine Calvin's parents and teacher's interactions with their own parents and teachers as they learned to count. Connections are made again and again to individuals and communities throughout history back to our human origins of counting. As Calvin counted, he brought forth a human history of counting in that moment. The history of mathematics is drawn into every mathematical act.[3] Therefore, there is potential for the conceptual reorganization and extension of networks within the historical and living "body" of mathematics to occur through any mathematical act. The potential and expectation for change emphasizes the ever evolving characteristic of mathematics.

Mathematics from this perspective suggests the mathematics within Calvin and Jolene's activities was not simply pulled out of them, as though their bodies or minds were containers of fully or partially formed mathematical concepts, nor was the mathematics pulled from an external environment to be used "as is." Aspects of the "body" of mathematics were pulled into action through Jolene and Calvin's own actions and interactions with one another and with the environment, and the mathematics was utilized to address their specific and immediate concerns. As the body of mathematics is pulled to action it potentially responds to their actions. Yet, the only aspects of the body of mathematics brought forth in each moment were aspects that Jolene and Calvin's own histories of experience would allow.

If we return to Kylie and Tamera's activities in chapter 5, we remember that Kylie and Tamera brought extensive histories of looking for and explaining number patterns within their mathematical actions. For example, their condition statements for any 2 x *n* were written as: "If *n* is even = *n*; if *n* is odd = *n* + 1." Aspects of the history of mathematics and their own personal histories are brought forth and drawn into these statements, which include algebraic functions and number theory related to odd and even numbers. Yet, if we showed Calvin and Jolene the condition statements for any 2 x *n* rectangle, they may not have been able to make use of these mathematical statements for future oriented actions. These condition statements evoke algebraic and function aspects of the "body" of mathematics that Calvin and Jolene's bodily histories of experience may not have allowed—at least, not in ways similar to Kylie and Tamera's activities.

As a person engages and responds to actions and interactions with another and with the environment, there is potential for conceptual reorganization of aspects of a person's own bodily networks of knowing within any mathematical act. The potential and expectation for change emphasizes the human capacity to learn. The intertwining of the history of mathematics brought forth through each person's history of experience lays a path of mathematical activity.

Laying Down a Path

> What we do is what we know, and ours is but one of many possible worlds. It is not a mirroring of the world, but the *laying down* of a world. (Varela 1987, 62)

The mathematical world brought forth through Jolene and Calvin's conversation is strikingly different from the mathematical world brought forth by Tamera and Kylie in chapter 5. Although both pairs share a similar concern for drawing a straight and accurate diagonal line, the actions which led to this concern and the mathematics brought forth to address it were different for each pair. Kylie and Tamera utilized the mathematics of common factors to address their concern for "where the dots go," while Calvin and Jolene took a geometric approach arising from Calvin's physical difficulties drawing the diagonal line. While Calvin's interaction with the physical environment had implications for the mathematical actions taken, Calvin's physical difficulties did not "cause" particular actions to occur, nor did his difficulties "direct" the path of mathematical activity. All actions occur in the space of interaction between persons and between persons and the environment. Calvin's difficulty drawing a straight line was not contained within his body or physiology. The act of drawing irregular lines occurred through the interaction of Calvin's physiology with the tools available to him. If a different tool was available, drawing a straight line may not have been a concern; if Calvin and Jolene had initially decided that the Diagonal Intruder entered the rooms on either side of the corner walls, the diagonal could have been drawn with perhaps a felt marker and the accuracy of the line would not have been as necessary; or if Jolene took on the responsibility for drawing the diagonal lines, the difficulty may also have been relieved. It is not surprising that Jolene did not take over the task of drawing the diagonal lines. The interpersonal relationship between Calvin and Jolene arose through a history of recurrent interactions with one another in this extracurricular program, but also through recurrent interactions as mother and son in nonmathematical environments. In this setting, drawing, folding, cutting, and building were generally roles fulfilled by Calvin, while Jolene generally fulfilled the role of a record keeper. The concern for drawing a straight line arose through the interactions among Calvin, Jolene, and the Diagonal Intruder prompt, between Calvin and the physical medium, and between the interpersonal roles and relationships enacted in this setting.

Viewing each action as an interaction between persons and environment suggests that an action or a concern does not belong to any one person in the conversation. Each action arises from physical and interpersonal interactions which connect it to a history of interactions within the immediate setting and back through to a distant past. As no action belongs to an individual, the path of mathematical activity in conversation cannot be directed by any individual. The path is laid on a moment-to-moment basis responding to the interactions of a multitude of sources. Actions or utterances do not or cannot specify a particular future or outcome. The pedagogical difficulty of specifying a particular path in conversation (or otherwise) is apparent at the onset of the narrative. Jolene initially turned to me to receive "direction." My suggestion was, "Just try different rectangles. Start with some simple ones like squares." This suggestion was part of the body of mathematics that I had brought to this setting. My immediate conjecture and expectation was that there may be a relationship between the diagonals of similar rectangles. For example, for any $n \times n$ rectangle the diagonal would go through n squares. My suggestion reflected *my own* history of experience in interaction with the prompt given.[4] The suggestion to look at squares was incorporated into the conversation, but in a way I had not anticipated. The first square Calvin drew was a 2 x 2 square, as opposed to the simplest of squares—a 1 x 1. The next drawing was no longer a square but a 2 x 3 rectangle. Although I *expected* a particular path of activities, my "directions" within the conversation did not specify a particular path. Although Calvin's 2 x 3 rectangle changed the path I had anticipated, it was one of many viable paths. From this perspective, actions and utterances from teachers or students are not expected to specify future action; however, they do open possibilities for future actions. The inability for any one person to direct the path of mathematical activity in conversation is not a limitation of mathematics instruction. It is simply a condition of the incompleteness of language and the incompleteness of our human existence (Maturana 1998). The path laid is one that is unique to the mathematical and interpersonal interactions taking place. The body of mathematics brought forth and the path laid through Jolene, Calvin, and my actions and utterances was indeed one of many possible worlds.

Narrative: Symmetric Properties of the Diagonal Line

Calvin began his fourth rectangle below his first 2 x 2 square. This time he outlined a 3 x 5 rectangle. He looked at it for several seconds.

I asked, "Where's the center of this one?"

Calvin didn't respond.

"What's half of 5?"

"Uh, you can't do half of 5."

"You can't?"

"No. It doesn't go in evenly."

"Oh, okay. What if you just guessed where the center was, how could you tell if you were right or wrong?"

Without speaking Calvin placed a dot near the center of the rectangle (see the dot on the left in Figure 24). He counted the squares on the left of the dot "1, 2." Then he counted the ones on the right, "1, 2, 3." He erased his point, put a new one to the right of it in the middle of the square, and counted again. "1, 2," counting full squares to the left. "1, 2," counting full squares to the right. "There." He drew a freehand line from the top left corner to the center then stopped. Then he turned the

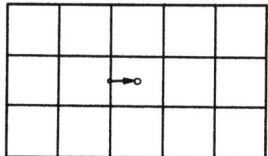

Figure 24. Finding the center of a 3 x 5 rectangle

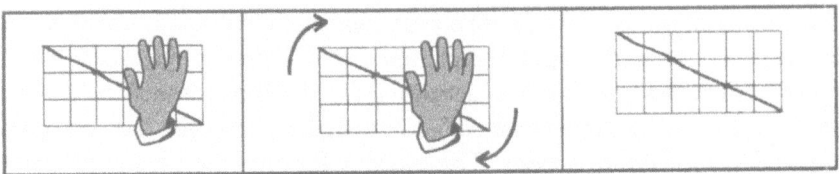

Figure 25. Rotational symmetry

page upside down and drew from the vertex to the center on the other side. After rotating his page back to its original position he counted the rooms entered, "1, 2, 3, 4, 5, 6, 7. 7."

"7? Okay." Jolene recorded: 2 x 5 = 7. "Oh, you did 3." She erased the 2 from her equation so that it read 3 x 5 = 7.

Again moving to the right on his page, Calvin drew another rectangle. This time, a 3 x 6 rectangle. He drew a point in the center, counted three full squares on either side and then drew a diagonal to the center from the top left vertex; turned his page around and drew the other half of the diagonal from the vertex to the center. "1, 2, 3, 4, 5—uh, is that 5 or nothing?" pointing to a possible point of intersection between the diagonal line and the grid.

"Let's see. I don't know." They hesitated for several seconds.

Suddenly shifting my orientation to the geometric properties of the rectangle, I pointed to the 3 x 6 diagram and said, "If the diagonal goes through up here, then it has to go through this one here, too. "Cause this," covering up the right half of the rectangle, "is the same as this," turning the sheet around and covering up the other half (see Figure 25).

"Oh yah. Do we have any others that do that?" Jolene asked.

"I don't see any," I responded.

Jolene looked at their previous diagrams and said, "This is the first time he's gone through a side wall" as opposed to the center of the rectangle. "What's one that would do that?"

I suggested, "You could try a 4 x 4. I know that one would."

Calvin drew a 4 x 4 rectangle and a diagonal line (see Figure 26). Jolene covered up the right side of the rectangle, "See, it goes through there." She turned the sheet around covering up the other side, which was still the right side of the rectangle, "and it goes through there, too." Jolene recorded 4 x 4 = 4 on her sheet.

For the remainder of the session they used a guess and test strategy to find rectangles that had the property where the diagonal intersected the grid in the top left quadrant as well as a bottom right quadrant. By the

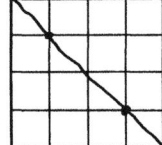

Figure 26. 4 x 4 rectangle

time the class ended, approximately one hour after beginning to work on the Diagonal Intruder, Calvin and Jolene found four rectangles with the symmetric property they were interested in: 3 x 6, 4 x 4, 3 x 9 and 4 x 8. Jolene continued to keep records throughout their investigation, but they were referred to only on occasion.

Discussion of Symmetric Properties Narrative

Lingering

> In a wood, you go for a walk. If you're not forced to leave it in a hurry to get away from the wolf or the ogre, it is lovely to *linger*, to watch the beams of sunlight play among the trees and fleck the glades, to examine the moss, the mushrooms, the plants in the undergrowth. (Eco 1994, 50)

Calvin and Jolene's activities could be dissected and described as a discrete set of obstacles addressed and overcome linearly; however, this perspective tends to sever the connections between their actions and interactions. An alternative is to view their mathematical activity as an ongoing act of lingering. From this perspective, a mathematical path is not viewed as being directed towards a future goal or as an attempt to solve a predetermined problem. Lingering is used here to describe a mode of being or living in which questions and concerns are raised and addressed in the moment as part of the persons ongoing interactions with each other and with the environment. Such concerns cannot be viewed as discrete, nor do they ever appear to be completely resolved; rather, they overlap, are revisited, and lead to other concerns.

The first section of narrative closed with Jolene's statement that the diagonal line had to go through the center of the rectangle. This utterance arose from previous actions, including Calvin and Jolene's concern for drawing a straight line. My question to Calvin at the onset of the second section of narrative, "Where's the center of this one?" draws upon Jolene's utterance. Calvin did not respond to this question, so I posed another question: "What's half of 5?" Calvin's response was not the "one" I expected. His answer, "You can't do half of 5. . . . It doesn't go in evenly," appears unrelated to the actions within this setting. My question, which abstracts the number 5 from the length of the rectangle and imposes an image of fractions on the diagram, may not have made sense in relation to the 3 x 5 rectangle in front of Calvin. His response perhaps drew upon his experiences in previous mathematical settings.[5] Calvin's history of experience did not allow him in this instance to bring forth aspects of the body of mathematics involving fractions as abstract quantities. It was only when my question was reframed within the present setting, "guess where the center is," did it provide an opening for him to act.

Locating the center point became part of Calvin's actions throughout the remainder of the session. While this action was used as a means through which to draw the diagonal it had implications for future actions. His procedure for drawing the diagonal line from the top left vertex to the center and then rotating his page and repeating that procedure later became an action used for explaining and locating points of rotational symmetry along the diagonal. That is, the right half of the rectangle was covered and the left half was examined, then the page was then rotated 180° and the other half of the diagonal was revealed as similar (see Figure 25). The rotational symmetry was another action that arose in response to their concern for drawing a straight line, but it had implications for future actions as well. The rotational symmetry was of aesthetic interest to Calvin and Jolene and their awareness of it led to a new concern. They began distinguishing rectangles which had the geometric property in which the diagonal line intersected points on the grid in places other than the center. We might anticipate that given a longer period in which to linger, Calvin and Jolene may have begun to make conjectures as to what rectangles would have this property. Or it is possible that a new concern may have been raised through their interactions and their path would have turned in another direction.

Describing Calvin and Jolene's activities as lingering emphasizes the interconnections between actions. Addressing one concern creates opportunities for new concerns. As ideas and concerns arise in the course of action, they are implicated in future actions. Questions, ideas, and resolutions are continually reinterpreted and folded into new ideas and actions. Concerns are revisited and relived, yet in being relived they are extended, elaborated, and redefined. A mathematical conversation sponsors and is characterized by these acts of lingering. It is through lingering that Calvin and Jolene maintain their relationships with one another and with the mathematics brought forth.

Opportunities to linger address a concern that may be shared by many mathematics educators; that is, as students strive to reach a prespecified goal or outcome they may miss the beauty of the "woods" or of the mathematics itself as they race through to a final destination. What if the goal for activity was not to enter the woods and strive to find an escape out of it? What if the intention was not to exit the other side at all? What role could educators play in helping students lay paths that allowed them to linger; that allowed them to maintain their relationship with the mathematics for extended periods of time? Lingering provides an alternative vision of mathematical activity or mathematical practice that emphasizes ongoing reflective activity.

Yet, lingering, as less goal-directed mathematical activity, may be viewed with skepticism. Although Calvin and Jolene engaged in mathematical actions for an extended period of time, would such mathematical lingering be observed as intelligent behavior or as random wanderings? Prior to making such a decision we need first to consider what is meant by intelligence.

Defining Intelligence

In his book, *The Society of Mind*, Minsky (1985) wrote of the difficulty of defining intelligence:

> CRITIC: How can we be sure that things like plants and stones, or storms and streams, are not intelligent in ways that we have not yet conceived?
>
> It doesn't seem a good idea to use the same word for different things, unless one has in mind important ways in which they are the same. Plants and streams don't seem very good at solving the kinds of problems we regard as needing intelligence.
>
> CRITIC: What's so special about solving problems? And why don't you define "intelligence" precisely, so that we can agree on what we're discussing?
>
> That isn't a good idea, either. An author's job is using words the ways other people do, not telling others how to use them. In the few places that the word "intelligence" appears in this book, it merely means what people usually mean—the ability to solve hard problems. (71)

If the definition of intelligence is the ability to solve hard problems, then Calvin and Jolene's acts of lingering above would likely not be judged as intelligent. Maturana and Varela (1987; Varela, Thompson, and Rosch 1991), however, present an alternative description of intelligence. They view the processes of evolution and cognition as "flip sides of the same conceptual coin" (Varela 1987, 49). From this perspective, intelligence for all organisms can be observed or judged using the same evolutionary criteria. Rather than observing the process of intelligence and evolution as "survival of the fittest," Varela, Thompson and Rosch (1991) suggest that we view intelligence and evolution as "survival of the fit." That is, any path or adaptation which allows an organism to maintain its relationship with its "medium" in a continuous history without disintegration may be viewed as intelligent action. Calvin and Jolene's path of mathematical activity cannot be viewed as random, as their histories of experience and the mathematics drawn into the moment do not allow just any path to be laid. Within this perspective, Calvin and Jolene's actions need not be judged as "right" or "best," only whether they were "good enough" to allow for a continuous relationship with one another and with the medium.

This view of intelligent action is possible only when we remove the frames of time and certainty around our actions. We often judge our discrete actions according to goal-oriented efficiency and attainability. These judgments are only possible if we perceive that we are capable of stepping outside the happenings of our lives. Removing the frames of time and certainty allows us to recognize that an organism's actions, including our own human actions, over a lifetime and over many lifetimes, arise from a multitude of interactions. We exist within a mode of lingering. The goal of living is not to find one's way to the other side of life, but to experience living through a series of meaning-making actions—actions through which "one's world stands forth" (Johnson 1987, 175).

Choosing to view mathematical cognition as a process of ongoing, successful living guided by sense-making actions allows researchers, teachers, and students to observe purposeful mathematical activity through lingering, rather than solely as finite problem solving. Within a mathematical conversation, as long as our actions are "good enough" to allow us to maintain our relationships with one another and with the mathematics without disintegration then they may be viewed as intelligent. From this perspective, "intelligence shifts from being the capacity to solve a problem to the capacity to enter into a shared world of significance" (Varela, Thompson, and Rosch 1991, 207). If the actions of Calvin and Jolene are judged under this definition, then their actions are observed to be intelligent. Their relationships with one another and with the mathematical environment were maintained as they brought forth a shared world of significance. We do not have to view Calvin and Jolene's activity surrounding the Diagonal Intruder as a separate event aimed at solving a problem, but rather as a moment in a continuous and infinite conversation with a mathematical world they brought forth.

Mathematical Practice and Game Playing

> There are at least two kinds of games. One could be called finite, the other infinite. A finite game is played for the purpose of winning, an infinite game for the purpose of continuing the play. (Carse 1986, 3)

Mathematical practice may be viewed from within Carse's (1986) framework of finite games and infinite games. These games also provide a means to further understand the perspectives that intelligence or cognition may be viewed as either finite problem solving or as ongoing, successful living. Viewing intelligence through problem solving or adequate action is a choice made arising from a person's emotional predisposition in the moment of action. Different emotions give rise to different "games" being played or to different mathematical practices. What is said and thought, the flow of actions, the relational interactions amongst the "players," the constraints on ways of knowing and thinking, how the "course is run," and what reality is brought forth are all implicated in the game played and the discourse within which the game is played.

As observers, we can examine Calvin and Jolene's activities for acts of problem solving. In doing so we can distinguish over ten problems that they solved, including the number of rooms the Diagonal Intruder entered for a 2 x 2 hotel, a 2 x 3 hotel, a 2 x 4 hotel and so on. That is, each rectangle created could be viewed as a distinct problem, just as a series of questions in a mathematics text is often viewed by students and teachers as separate activities. Although we can assume that learning occurred throughout their engagement in these related problems, in our observations we can make distinctions between their activities and view Calvin and Jolene engaging in a number of independent finite games. From another perspective, and as has been stated earlier, these "problems" arose as part of Calvin and Jolene's ongoing activity and are necessarily interdependent; the choice of rectangle or problem was made in relation to their immediate concern, and the means through which a solution was determined for a rectangle often gave rise to other concerns and other actions. From this perspective their activity may be viewed as an infinite game. Finite and infinite games are not dichotomous positions. Rather, an infinite game may encompass a number of finite games (although the reverse is not possible). The problems or finite games that Calvin and Jolene played were part of an ongoing infinite game they were enacting. A distinction between mathematical practice as a finite game and as an infinite game provides two ways of observing other's activities, and also two ways in which a person can engage in and observe his or her own mathematical activity.

Mathematical Practice as a Finite Game

When a person views the practice of mathematics as a finite game, as a series of discrete tasks, problems and truths, every move made is for the purpose of winning—of bringing the game to an end and being released of the relationship between oneself and the mathematics. When one plays mathematics as a finite game he or she solves problems in order to complete them and put them aside. Each problem solved is viewed as a victory. The desire to win—the desire for certainty and closure—is the emotional basis giving rise to mathematics played as a finite game. A person expects problems to be challenging and difficult, but once a solution is determined "it is as it is"—the quest for certainty has been achieved.

When mathematical practice is viewed as a finite game, every action that is not tied to the goal of winning or completing the task is not part of the game and may even be seen as a nuisance, a source of frustration, or as a waste of time. Time in finite games is viewed as a commodity, and there is a pressure or need to get someplace, to arrive somewhere (Davis 1994). The unpredictable and uncontrollable are dreaded, for they are obstacles in the path towards certainty, towards knowledge.[6] A mathematical problem viewed as a finite game is played within predefined boundaries. The rules for play within these boundaries remain consistent. If the boundaries or rules change, a different game is being played, a different problem is being solved. Attempting to change boundaries in the midst of play may create a great deal of frustration for a person either observing or playing mathematics as a finite game. Changing the boundaries or manipulating the rules to accommodate certain actions or solutions may appear as cheating; or, activity outside the boundaries may be viewed as a diversion or as off-task or off-track behavior because it is at odds with the purpose for engaging in mathematics. Towards the end of Calvin and Jolene's activities they become interested in the rotational symmetry of the rectangles and they spent over half an hour finding rectangles which had the property whereby the diagonal line intersected the grid in places other than the center point. A person observing their activities from the perspective of mathematics as a finite game may see this as an off-task or an unproductive diversion. A teacher, viewing the Diagonal Intruder as a problem bounded within the mathematics of number patterns and function relationships, may attempt to pull Calvin and Jolene "back on track" by suggesting that they continue to look at the rectangles systematically.

The quest for certainty is an important aspect of mathematical practice. Viewing and engaging in "mathematics as problem solving" and "mathematics as proof" satisfy the emotional desire for certainty. The desire for certainty gives rise to the practice of mathematics viewed as a finite game. Such games need clear boundaries and observable endings. When mathematics is played or observed as a finite game, one's progress towards certainty or terminal understanding becomes important. Within this perspective, the body of mathematics often arises and is perceived as a static entity. Judgments of intelligence or achievement can be measured by how close a person is to displaying the expected understanding or how well a person's actions match what may be deemed as "correct" mathematics.

The emotions underlying mathematical practice as a finite game are perhaps best satisfied within an argumentative discourse. All mathematical actions and explanations within this discourse are aimed at progressing towards a common goal. Finding errors and counterexamples, and defining clear boundaries to a problem are important moves towards determining an ultimate answer or explanation for a prespecified goal. When its participants "arrive" at a solution or explanation, there is a sense of certainty, and there is satisfaction in obtaining that certainty. Certainty is an emotional rather than an actual requirement. Many people realize, even as they play mathematics as a finite game, that solutions or proofs may later be judged inadequate or faulty; however, this does not diminish the feeling that the knowledge or understanding that has been obtained in that moment is somehow *real*.

Mathematical Practice as an Infinite Game

When mathematical practice is viewed as an infinite game, the desire for certainty is relinquished and the need for closure is not felt. Acceptance of uncertainty and a desire to prevent the game or activity from ending are the emotional bases of mathematics played as an infinite game. Rather than achieving a state of certainty, Calvin and Jolene's actions in this chapter and Kylie and Tamera's as well as Stacey and Ken's actions in the previous chapters often brought about further uncertainty and surprise. Players engage in infinite mathematical activity with an openness to the unexpected, and they resist "the temptation of certainty" (Maturana 1998). A person who views mathematics as an infinite game never "feels" that his or her activity or knowing within a mathematical situation is ever final or complete.

It is generally not useful to view mathematical activity from this perspective, using a metaphor of movement or "progress" towards a particular destination such as a specific goal or outcome. While Jolene and Calvin's actions may appear to anticipate a future goal at any moment, it is difficult to view them as making progress towards a known goal because as their questions and concerns change, the goal anticipated changes. They may never reach the expected destination. For example, when Calvin and Jolene's understanding of the Diagonal Intruder suddenly broadened to include an awareness of the rotational symmetry within the rectangles, the world within which they lived suddenly opened up. This understanding shifted the boundaries of their activities and provided Calvin and Jolene with an opening to engage in an extended investigation of rectangles with a particular geometric property. The shift in boundaries around the problem, which may have appeared irrelevant if their activity was viewed from within a finite game perspective, is consistent with mathematics played as an infinite game. In fact, the boundaries and the rules for infinite play must change "to prevent it from coming to an end, to keep everyone in play" (Carse 1986, 8). Although Calvin and Jolene played within the mathematical boundaries they set for themselves, at any moment new actions and understandings were potential occasions within which to stretch the boundaries of activity so that interesting actions and understandings could be incorporated into their play. Movement within mathematics played as an infinite game is not towards a destination or a terminal understanding; it is movement which stretches boundaries to include new actions, new interests, and new concerns.

The desire to maintain a relationship within a mathematical setting and the acceptance of surprise and uncertainty are emotions giving rise to mathematical practice as an infinite game. Such games have flexible boundaries, and the boundaries within which one plays are determined by the mathematics and personal histories drawn into that moment. When mathematics is played or observed as an infinite game, all play exists in the now. It is not directed towards a prespecified goal, so it is impossible to measure how far a person is from a goal. We can, however, observe what mathematics is drawn into action in relation to the concerns the participants themselves pose; we can witness how the participants' actions broaden their understanding of the mathematical world they bring forth; and we can judge whether their actions are adequate by whether they have maintained their interaction with the mathematical world. Mathematics as an infinite game is not played within time constraints, as the game does not have a beginning or ending.

Mathematical practice as an infinite game is perhaps best satisfied within a discourse of mathematical conversation. Actions and explanations within this discourse allow its participants to maintain their relationships and stretch the boundaries so that new ideas, concerns, and understandings can be incorporated. A complete understanding of a problem is never obtained, but one's understanding continues to expand as he or she maintains relationships with the mathematical world brought forth.

Infinite play or lingering in a mathematical space is invisible to an observer who views mathematics as a finite game. "Such viewers are looking for closure, for the ways in which players can bring matters to a conclusion and finish whatever remains unfinished" (Carse 1986, 115). A teacher observing Calvin and Jolene's activities from the perspective of mathematics as a finite game may be somewhat frustrated with their actions. Over the course of an hour, Calvin and Jolene spent an inordinate amount of time attempting to draw straight lines, they did not appear to reach any conclusions, nor did they appear to be progressing towards a certain goal. From this perspective, their acts of lingering may appear as random wanderings. "If, however, the observers see the poiesis in the work they cease at once being observers. They find themselves in its time, aware that it remains unfinished" (115). It is at this point that observers or teachers may enter into the infinite game with their students, in which they no longer have to stand aside and "direct" activity but may bring their own histories of experience into the conversation and participate in the infinite game.

Notes

1. This program was organized and instructed by Professor Elaine Simmt, Department of Secondary Education, University of Alberta. See Appendix A for more information.

2. The reader may remember that this is the same prompt given to Kylie and Tamera and discussed in chapter 5. Chronologically, Calvin and Jolene's activity occurred prior to Kylie and Tamera's. This was my first experience with the prompt.

3. The use of the word "drawn" indicates the act of pulling mathematics into the moment, and the image that the history of mathematics is visually sketched into the moment as well.

4. My own work arose through actions involving similar rectangles and ratio which led to the statement: If a and b represent the two dimensions of a rectangle, the number of rooms entered is $(b + a) - G.C.F.$ [Greatest Common Factor].

5. This is another example of how our words cannot specify an action or answer. Calvin's answer is not incorrect. In fact, it is acceptable given the constraints he places on dividing numbers in half.

6. An example of mathematical practice as a finite game which demonstrates the emotional desire to "arrive" at a solution quickly and the frustration of obstacles and off-track behavior is available in Appendix B.

Chapter 8

Re-turning to Features of Mathematical Conversation

It is perhaps time to return to the point where we started; to revisit the purposes of this research. Throughout this work, I hoped to illustrate and articulate the living practices of persons engaged in mathematical conversation, knowing that such practices were not objects with properties, but events always in the moment of becoming. Thus, the "body" of mathematics was revealed not as a set of content, but as a responsive and evolving network. It is an explanatory domain allowing us to find meaning and coherence within the ongoing events of our lived experiences. The practice of mathematics, then, becomes a continuous meaning-making venture that may only be viewed on the surface as a utilitarian problem-solving pursuit, but is, at its depths, an emotional, embodied, and interactive endeavor in which our mathematical understanding stands forth as "a series of ongoing related meaning events" (Johnson 1987, 175). While I feel I have continued a conversation begun earlier by Lampert, and even earlier by Lakatos and Polya, or perhaps even earlier by Hilbert, I am at somewhat at a loss knowing where to begin this ending, or end this beginning, for writing a final chapter has consequences.

In this chapter I hope to summarize the features of mathematical conversation into a coherent whole. But if such features are placed into a noncontextual form, I risk all that I hoped to accomplish. Such propositions may take on the guise of universal truths in which mathematical conversations appear to exist as objects somehow independent of my own actions. Also, such a summary may appear to be posed as if to an audience behind glass; an audience who is not expected to respond. Such a re-presentation, offered as a conclusion, may bring the conversation to a close, when my intended purpose is to continue the conversation about mathematical discourse and practice by offering mathematical conversations as a possible alternative.

Rather than a conclusive ending, the statements I make here are a continued quest for meaning and coherence. Perhaps I can summarize this work in a manner whereby both I and the readers know that such a summary is an offering made still in the midst of understanding; as a way of helping me and the readers organize what has occurred thus far, while recognizing that what I offer will be incomplete. Incomplete in the sense that I cannot repeat all that has arisen in the past, and any re-presentation now is yet another interpretation, another reformulation of past events. Further interpretations and explanations will necessarily revise the present offering; indeed, such reinterpretations and revisions are essential for keeping the conversation going. Our reasons for becoming aware of alternative discourses and legitimate practices are not because we are striving to find a discourse that eventually "works" or a practice which prescribes what we should do as educators. Rather, continuing the conversation allows us to question our assumptions about how mathematical knowing is generated and accepted by the discipline and encourages us to return to and understand how these assumptions have been woven into classroom practice. What I offer is hopefully "good enough" in that it initiates new questions, expands our horizons for understanding mathematical practice and discourse, and allows us to continue to explore mathematical conversations and investigate alternative descriptions of mathematical practice.

Mathematical Conversations as a Gestural Genre

I suggested in chapter 4 that perhaps mathematical conversation is its own speech genre: a relatively stable and stylistic set of utterances defined by its conception of the addressee (Bakhtin 1986). Bakhtin asserts that from the first word, we listen for and speak into a particular genre. A genre is identified by its own thematic content, style for interaction, and narrative structure. In addition, a horizon of expectations, assumptions, and emotions is brought to bear on the genre (Bakhtin 1981). I propose at this time to return to the idea of a speech genre as a means for addressing and organizing the features of mathematical conversations. However, rather than a speech genre, I am choosing here to label it as a *gestural genre.* The word *gestural* is an adjective "designating or pertaining to the theory that human speech originated in oral imitations of bodily gestures" (O.E.D. 1992). Therefore, by emphasizing "gesture" rather than "speech" I am drawing attention to the notion that speech and language are the physical gestures embedded in the sensorial and sensual dimensions of experience (Abram 1996). A gestural genre rather than a speech genre is also particularly appropriate for mathematics, as much of the interaction that occurs simply cannot be understood through utterances alone. Participants in a mathematical conversation use and create a gestured language pertaining to their mathematical practice and consisting of a unique set of embodied and interactive gestures—verbal, physical, pictorial, and symbolic—to communicate and create aspects that the mathematical world brought forth.

The features of mathematical conversation as a particular gestural genre have been organized below into three gestural domains of discourse: *addressivity towards the other; addressivity towards otherness; and the lived curriculum.* The two domains of addressivity point to the interaction among persons and between persons and the mathematical environment; arising from these interactions a path of mathematical activity is laid which is and becomes the lived curriculum. It is also important to note that assumptions, expectations, and emotions intrinsic to mathematical conversations and mathematical interactions in general surface and are addressed throughout the three domains.

The gestural domains provide a framework for discussing mathematical conversations and for discussing other gestural genres that may occur in the mathematics classroom. Each domain is used as an observation tool for drawing attention to various aspects of mathematical interactions. Derived from the mathematical conversations illustrated in this monograph and in relation to this work, features of mathematical conversations have been identified and described within each gestural domain. A synoptic overview of the features of mathematical conversations as a gestural genre in relation to the three gestural domains is provided in Figure 27 (page 149) at the end of this chapter.

Addressivity Towards the Other

> An essential (constitutive) marker of the utterance is its quality of being directed to someone, its *addressivity*. (Bakhtin 1986, 95)

The nature of addressivity defines the style for interaction between persons and becomes a primary means for identifying the interaction as a particular gestural genre. Addressivity is the quality with which one turns to the other. It draws attention to the responses provided to and anticipated of the other, as well as the emotional basis of the relationship. Addressivity orients the observer's gaze towards all persons involved in the interaction and the relationship they have established amongst themselves. In any mathematical interaction we, as observers, can ask ourselves, "What purpose does the other appear to serve?" Or perhaps we might ask, "What form of response is anticipated or demanded of the other?" These questions may be asked as an outsider to the conversation, or asked of oneself as an insider, be it a student, a teacher, or a researcher. We would expect to answer these questions differently depending on the gestural genre we are observing.

In some interactions we can imagine ourselves as listeners in which the speaker continues to speak regardless of to whom he or she is speaking. As the other, we serve only as a body in the audience; it matters little who we are, as no response is anticipated. Listening to a classroom lecture or presentation without an opportunity for interaction is an example of addressivity in which the speaker may turn *at* the other, but does not alter his or her actions or utterances by responding *to* the other. There are also times when a similar addressivity occurs even in more intimate settings such as small group discussions or exchanges over coffee. Occasionally in our interactions with another we intuit that our gestures, or more significantly, we, as participants in the dialogue, are not accepted or even acknowledged by the other. A gestural genre in which the other is not acknowledged and no response is expected could be defined as a monologue:

> Monologism, at its extreme, denies the existence outside itself of another consciousness with equal rights and equal responsibilities, another *I* with equal rights (*thou*). With a monologic approach (in its extreme or pure form) another person remains wholly and merely an object of consciousness, and not another consciousness. No response is expected from it that could change everything in the world of my consciousness. . . . Monologue manages without the other, and therefore to some degree materializes all reality. Monologue pretends to be the *ultimate word*. (Bakhtin 1984, 293)

In an extreme form of argumentation, versions of dueling monologues may occur such that both parties speak without listening, neither party is open to change, and both parties hope to state the final word on the matter. However, this is generally not the case. In an argumentative genre the other serves a more visible and purposeful role than is implied in the above description of a monologue. An identifying characteristic of an argumentative discourse may be observed in the presentation of my mathematical argument: I *address* the other because I want to *convince* him or her of its truth. This gestural genre does not ignore, but demands a response from the listener. The response I expect and listen for in the other's gestures is either *agreement* or *disagreement* with what I have proposed. While a monologue continues regardless of whether the audience agrees with or understands what has been said, in an argumentative genre if both parties have presented their arguments, and if they agree with one another, the discussion ceases, at least on that topic. A discourse of argumentation flourishes in disagreement. The addressivity of argumentation may be characterized by the quality with which participants attempt to compel each other to change viewpoints as well as their listening for and making gestures of agreement or disagreement. Emotional satisfaction is achieved at times when I feel I have successfully presented the stronger, more rational argument, which at its best also includes changing the viewpoint of the other person. I part from such interactions with a sense of victory—I have won the argument.

By way of contrast, a conversation as a gestural genre is not a platform in which to enforce and maintain a personal standpoint; instead, it presents an opportunity for participants to bring ideas into play, thus, putting themselves at risk, and being changed, not by force, but by questioning their own assumptions and opening themselves to the experience (Gadamer 1989). As Grudin (1990) writes, "We cannot open ourselves to new insight without endangering the security of our prior assumptions. We cannot propose new ideas without risking disapproval and rejection" (9).

In Stacey and Ken's activities in chapter 6, it was noted that in conversation an intimate space is created in which both participants are fully present and fully responsible for sustaining and constituting the world brought forth. The participants' actions together are fully implicated in what is said and done. The addressivity within a mathematical conversation is characterized by the quality of conversing with, of turning with, another to generate explanations for the questions that arise through their experiences together. In a mathematical conversation, I *address* the other because I want to share my present mathematical *understanding* with him or her about a mutual topic of concern, and I request that the other's response continue and perhaps extend my understanding of this topic.

Allowing one's understanding to change and expand in conversation, a characteristic of its addressivity, is an emotional choice that one makes. Change does not occur through force of reason, but by choosing an emotional disposition that allows oneself to be "moved" by the gestures of the other. Similarly, the choice to be closed to the ideas of the other is also emotionally based. The fundamental basis of addressivity occurring in every form of gestural genre, in every form of interaction, is emotional. In conversation the emotional choice made is one of mutual acceptance or love of the other. It is only through love that a social phenomenon is possible (Maturana and Varela 1987).

We do not exist and, therefore, cannot interact with others as fully rational individuals. Our actions and interactions, even (and perhaps especially) in mathematics are always emotionally based. Differences and disagreements often reveal the emotional choice of acceptance or negation of the other within the interaction. It may not be until after differences arise that we recognize whether or not we have been accepted, or even whether we have accepted the other person as a partner in the present dimension of living (Maturana and Varela 1980). When disagreements do arise, we cannot be forced by appealing to reason to accept the other's viewpoint. Willingness to change one's point of view is based on an emotional willingness to be open to the other's viewpoint and to view the other's gestures as having worth. Mutual acceptance allows us to remain open in times of disagreement and continue to let the other speak to us.

> In human relations the important thing is. . .to experience the Thou truly as a Thou—i.e., not to overlook his claim but to let him really say something to us. Here is where openness belongs. But ultimately this openness does not exist only for the person who speaks; rather, anyone who listens is fundamentally open. Without such openness to one another there is no genuine human bond. . . . Openness to the other, then, involves recognizing that I myself must accept some things that are against me, even though no one else forces me to do so. (Gadamer 1989, 361)

If the mathematical relationship is based on negation, "things that are against me" may be perceived as threats. In these instances I try to *defend* my own position rather than open myself to the position of the other. Conceptual change does not occur if both parties take a defensive stance and try to compel the other to change. On the other hand, in an interaction based on mutual acceptance, I am willing to believe that the other person's explanations have worth, even if they go against me and differ from my own. Therefore, a similar disagreement may arise in two different mathematical interactions, but depending on the emotional basis of the interaction it may be perceived and acted on very differently. In a mathematical conversation which brings forth a conversarial reality, differences are not simply tolerated or ignored, nor are they simply agreed or disagreed with. Differences and disagreements are viewed as invitations for dialogue (Maturana 1988). Disagreements present opportunities to gain awareness of our prejudices and broaden our mathematical understandings of the world we bring forth.

In our conversations we create a space for each other that allows me to see the other as working "beside" me on the same project. We turn with one another to broaden our understanding of a mutual concern. In conversation, "one should address others with a presumption that they are capable of responding meaningfully, responsibly, and, above all, *unexpectedly*" (Morson 1981, ix). The unexpected in mathematical interactions may occur as an alternative explanation or a counterexample which goes against a previously accepted explanation. Counterexamples and alternative explanations which arise and are sought in the course of our conversations are not "aimed" at me to prove me wrong or to force me to change. They are not perceived threats against me. The counterexamples and conflicting explanations are gestures offered as a means to broaden our understanding of the mathematical world we bring forth. Explanations, even when they differ from my own, are offered in the flow of understanding gestures as extensions of what has been previously offered and experienced. It is my *responsibility* in conversation to try to understand the other's explanation and offer an understanding gesture in return. The sense of responsibility is an ethical imperative implied in the addressivity within mathematical conversations: a person is responsive to, responsible for, and respectful of oneself and the other. Such mutual acceptance is the "basic stabilizing factor in the constitution" of a mathematical conversation (Maturana and Varela 1980, xxvi). Satisfaction within the conversation does not occur when I have presented a strong argument, or convinced the other to accept my viewpoint. It occurs in the perception that I have broadened my understanding through the experience.

Pedagogical Consequences of Addressivity towards the Other

The emphasis on communication in current mathematical reforms and its potential to promote learning has encouraged educators to provide more opportunities for whole-class discussions, small group work or paired interactions. It is important that the teacher consider the level of intimacy possible within each type of interaction and the implications that the physical and emotional space created has on an individual's opportunity for communicating and constituting the mathematical phenomenon brought forth. An awareness of addressivity draws attention to the quality and emotional basis with which one turns to another within these physical and emotional spaces. When a student speaks, what purpose do the other students or the teacher appear to serve? What form of response is expected and provided by these others? Relationships based on an emotion of mutual acceptance or love may be observed in the flow of understanding gestures that occur, particularly in instances of difference. The listeners would be expected to ask genuine questions for clarification or offer continuations of the gesture. Counterexamples become important to the conversation, not as means to refute the other, or even the other's explanation; counterexamples allow the participants to question the assumptions on which the explanations were previously made and open the possibility for further action and a broader understanding.

An understanding of the emotional basis of interaction has implications for orchestrating discourse within the classroom. Mathematics learning cannot be promoted as simply a rational enterprise. If we expect students to be willing to share their own explanations, to be open to other viewpoints, and to try to understand explanations which differ from their own, a community based on mutual acceptance or love must be established. The pedagogical practice of providing opportunities for students to listen to one another re-present thinking using convincing arguments, and encouraging students to agree or disagree with the solutions offered, is not sufficient for expanding understanding if students have not chosen mutual acceptance as their emotional predisposition within the interaction. Such a practice may even encourage students to maintain and defend their own viewpoint by looking only for faults in the other's reasoning, not for ways in which they can learn from the explanation. This practice may unintentionally foster a climate of negation. That is, if a student believes in the certainty of his or her own answer, classmates serve no purpose other than to be convinced or converted. At these times, there is no need to listen to other explanations with a purpose of understanding. In the classroom we cannot expect that all students will immediately choose an emotional predisposition of mutual acceptance; however, it is likely that this choice gains appeal through recurrent interactions in a responsive and respectful environment, whether that environment is mathematical or otherwise.

There is potential for the teacher to enter into conversation with his or her students as a temporary or periphery member, but addressivity concerns must be considered; that is, the teacher must also exist in a relationship of mutual acceptance with similar expectations for responding to and acting with the students in conversation. The teacher must be invited into conversation and must expect that his or her participation in different conversations will be uniquely determined by the participants themselves. He or she cannot expect to direct activity, but should see conversation as an opportunity to ask questions arising from actions, share experiences in relation to the interactions, draw upon aspects of the body of mathematics that expand the students' understanding of a mathematical phenomenon brought forth, and be open for his or her own mathematical understanding to be expanded through unanticipated interactions with the students. There are, of course, barriers to the teacher's participation in conversation. The most prominent may be the history of recurrent interactions students bring with them from other communities in which the teacher has generally appeared to possess answers to any and all mathematical problems. When this relational expectation is in place, the presence of the teacher may shift the reality brought forth from a conversarial reality of coemerging actions and explanations, to a universal reality in which answers are perceived to preexist actions and explanations.

An important question arising from this work is how is it possible to establish and maintain a mathematical community including both the teacher and the students based on mutual acceptance whereby the members of the community feel responsible for attempting to understand the explanations of others? Does providing recurrent opportunities for mathematical conversations help establish and support a community based on mutual acceptance?

Addressivity Towards Otherness

A gestural genre is defined by its addressivity towards the other and perhaps also towards otherness. Rather than directing attention towards the relationship between persons in interaction, *addressivity towards otherness* orients the observer's gaze to the relationship between the persons involved and the mathematical world they bring forth. The observer might ask, "What form of response is anticipated or demanded of persons by the otherness or of otherness by the persons involved?" Or, "What purpose does 'it' serve?" Once again, we may attend to the emotional basis of this relationship and the rhythm of mathematical interactions that occur.

In a traditional mathematics classroom "it" may be a mathematical question, a symbolic gesture on a worksheet. As a student interacting with the worksheet, I address the question posed and it addresses me. The gestured response it demands of me is a symbolic answer. It matters little what the answer is, how I determine it, or even who I am. My emotional response to it may be one of obedience, in which I try to answer correctly, or my response may be of indifference, in which I may answer, but am apathetic to whether it is correct or not. When I respond "correctly"—that is, when my response *matches* the expected answer—my answer becomes nameless and faceless, indistinguishable from other correct answers. The question is not altered in any way by my gestured response. Similarly, I may be seldom or perhaps only slightly altered emotionally or cognitively by responding to it. If the relationship is based on tolerance, other than a passing acknowledgment, I rarely let the question address me or speak to me. As soon as I have provided an answer, the relationship ends and I move on to the next question. Although I have emphasized emotions of obedience and indifference, for some students, including myself as I alluded to in chapter 1, the questions may be viewed as a challenge, as finite games to be played. In these instances, a relationship of rivalry, discussed below, may arise.

In a mathematics classroom based on constructivist principles, otherness is more likely to be a predefined problem or activity with assorted manipulatives or mathematical tools available. The problem is also unlikely to belong to anyone, only to the perceived realm of mathematics. A characteristic of a problem is that I must *see* it as a problem and be motivated to solve it (Charles and Lester 1982). A problem is only defined as a problem when it addresses me. I *address* the problem as an *obstacle* which I am motivated to overcome. I respond to the problem by engaging in a gestural process through which I hope to break through the obstacle to determine a viable solution to the problem. My emotional predisposition for engaging in a relationship with the problem may be my desire to conquer it. In these instances, a relationship of *rivalry* is formed. Once I am able to defeat it, it no longer holds my interest and I am released from the emotional bond between us. I part from this interaction with a feeling that I have either defeated the obstacle or it has defeated me. I have either won or lost. When either I or otherness has been defeated, the relationship ends and I am free to move on to another problem.

As illustrated in chapters 5 to 7, in a mathematical conversation the participants are presented with a prompt, rather than a problem.[1] Assorted mathematical manipulatives and tools are available. Similar to the previous description of gestures, the tools available are implicated in the explanations provided. However, a unique feature of mathematical conversations as a gestural genre in relation to addressivity towards otherness is that both the participants in conversation and the mathematical otherness are addressed and both are expected to respond.

The mathematical world brought forth is constantly responding to the questions posed and explanations provided, which is the essence of what makes mathematics a human activity. Understanding of a mathematical world, then, is never finalized or complete. The participants are always responding and existing on the threshold of understanding. The continuous relationship between the participants and the mathematical world gives rise to new actions and experiences, new questions and concerns, and new explanations in a recursive cycle. The rhythm of interaction between others and otherness in a mathematical conversation is characterized by this recursiveness.

The emotional predisposition for engaging in a continuing conversation is an ongoing sense of *curiosity* and *uncertainty* the participants have of the mathematical world brought forth. The continuation of the conversation is in large measure dependent upon neither person feeling he or she fully understands the mathematical world or the other person's explanations of it. Both persons bring an attitude of curiosity which allows them to play in and play with the mathematical situation, thus continuing to bring forth a mathematical world or play-space. Satisfaction is achieved through play—by maintaining relationships indefinitely through a sense of curiosity and through the feeling that one's understanding is continually expanding through interactions between participants and the mathematical world brought forth.[2]

Pedagogical Implications of Addressivity Towards Otherness

An awareness of addressivity towards otherness draws attention to what students and teachers perceive as appropriate responses to mathematical questions, problems, or prompts. In large measure the expectations surrounding activity are based on an emotional predisposition towards otherness. Relationships between persons and the body of mathematics may be based on indifference, obedience, tolerance, rivalry, or curiosity. The perceived expectations for appropriate responses and the emotional basis formed has implications for the flow of gestured responses from the participants, the desire to maintain or escape from the mathematical relationship, and the perception of mathematical practice and its legitimate activities within the classroom.

An understanding of the emotional basis of interaction between persons and the body of mathematics has implications for developing alternative expectations for mathematical practice within the classroom. The current emphasis on mathematical problem solving has the potential to promote mathematical understanding and development; however, this position must also take into consideration emotional predispositions fostered in the students' mathematical interactions. An emotional relationship of rivalry promotes an image that problems are obstacles to defeat or be defeated by. Both wins and losses impact the student on a personal and emotional level. Students viewing the problems as rivals may attempt to escape from the relationship once victory is achieved or when defeat is imminent or is perceived to have occurred. Problems line up as potential opponents, rather than as opportunities to linger and understand the potential landscape of a prompt or problem. Thus, students may be less willing to reflect on their own and on other students' processes for solving the problem.

However, a teacher may decide that in some instances a relationship of rivalry is most useful for a particular situation. For example, when the response expected is a single correct answer, it may be useful to view the expected answer as matching an independent world, such as in the mastery of a mathematical concept, or the practice of an isolated skill. A series of questions or problems allows students to practice an algorithm such that it becomes mechanical or ritualized and, therefore, it can be applied quickly and efficiently as the situation warrants. However, in these circumstances the student must be able to approach the problems with the belief that he or she can eventually be victorious and not concede to defeat too quickly.

If we want to provide opportunities for students to engage in activities in which the students feel reluctant to be released from the emotional bond, such as ongoing projects, experiments, or investigations, or if we want to encourage students to raise questions of concern in relation to some activity or mathematical practice, as teachers we need to be explicit about the type of responses expected of them and foster relationships based on curiosity rather than obedience, indifference, or rivalry. Of course the question is, how might an emotional predisposition of curiosity be fostered? And will recurrent opportunities for mathematical conversations help establish and support mathematical curiosity?

The Lived Curriculum

Through interactions between the participants in conversation and between the participants and the mathematical otherness, a mathematical world is brought forth. In bringing forth this world, a path of mathematical activity is laid which is identified here as the *lived curriculum*. The lived curriculum is an expression of curriculum as the running of the course, rather than of the course to be run (Pinar and Grumet 1976). An awareness of the lived curriculum, arising through a process of interaction between others and otherness, directs attention to mechanisms that orient the participants' activities within an ever-changing mathematical world. The constraints placed on what path or paths are laid through the interactions have implications for how a mathematics curriculum "as lived" is defined and what activities and responses are perceived as mathematically acceptable. An observer may ask, who or what is responsible for directing the path of mathematical activity? What mechanisms set the boundaries of the lived curriculum? And what mathematical gestures are used in expressing and constituting the path of mathematical activity?

Directing the Course of the Lived Curriculum

Traditionally we have thought that the teacher was responsible for directing the lived curriculum by choosing the questions and problems and ensuring that students used specifically sequenced symbolic algorithms and techniques so that processes and outcomes matched the pregiven and externally determined solutions. Our current perception is that while the teacher is expected to provide suitable problems and activities for the students to engage in, the actual path laid is to be autonomously constructed by the students through heuristics and invented algorithms using concrete, pictorial, or symbolic mathematical representations. Yet, there is still an expectation in many activities selected for students that the outcome should eventually match one that is externally predetermined.

In a mathematical conversation, as we have observed in the work of participants contained herein, the students do not set out to find a solution to a pre-given task. Their ongoing sense of curiosity and their responses to the body of mathematics and its responses to the participants' actions continue to alter the mathematical world brought forth. The lived curriculum is bounded by what both the mathematical situation and the phenomenological histories of the participants will allow (Varela, Thompson, and Rosch 1991). Therefore, there is no optimal path, only possible, potential, and viable paths.

Questions that *address* the participants become *mutually and emotionally charged concerns*. A mutual topic of concern is an emotionally based issue which *orients* the direction of conversation. Neither the conversation's direction nor structure exists within the control of either participant. Yet, once a concern has been raised and the participants begin to act, there is structure and direction to their actions. The direction of the lived curriculum shifts or, rather, *drifts* on a moment-to-moment basis as participants address topics of mutual concern that arise.

Mathematical Explanations

Mathematical explanations are significant points along the path of the lived curriculum. How mathematical explanations are posed and how they are accepted define mathematics as a practice and as a discourse. Explanations in the course of our everyday activities are offerings to ourselves and others to make sense of our experiences. Yet, underlying how mathematical explanations are posed and accepted is a set of assumptions, expectations, and emotional predispositions which have evolved through historical networks of mathematical conversations.

An awareness of mathematical explanations draws attention to the mode in which explanations are offered and accepted, and to the perception students and teachers have of what is an acceptable explanation within the domain of mathematics. An observer may ask, "What purpose does the explanation appear to serve in the interaction?" Or, "What criteria are used for accepting explanations?" The answers to these questions are in large measure dependent upon the form of explanation that is offered.

In chapter 5 I distinguished between two forms of explanations: *Explanations in action* and *explanations as re-presentations*. Explanations occurring in a mathematical conversation are predominantly stated *in action*, in the midst of coming to an understanding of a mutual topic of concern. Explanations in action are made as offerings or as invitations to oneself and to others to explain issues of mutual concern arising from interpersonal and mathematical interactions. They are judged as acceptable within the conversation when they appear coherent with the participants' previous experiences and provide enough understanding of the immediate concern so the participants can continue on. The explanations are not optimal, only sufficient, believable, plausible, and good enough for the temporal now. These features are consistent with the features of addressivity within conversation. That is, they are stated within an emotional predisposition of mutual acceptance and arise out of a curiosity towards the mathematical world brought forth.

Pedagogical Consequences of the Lived Curriculum

> A curriculum that is viewed as a domain for conversational action is inherently interactional: it includes the content knowledge emphasized in older versions of curriculum but insists that such content is of interest for the conversation. . . . Content that does not invoke further conversation is of no interest; it is dead as well as deadly, certain to bring the curricular conversation to a halt rather than leading it forward. (Applebee 1994, 47)

An awareness of who or what directs the path of mathematical activity is indicative of the gestural genre. An understanding of this process has implications for the lived curriculum in mathematics. If we expect our students to engage in mathematical conversations, we have to allow the topics of concern arising from interactions among students, the teacher, the prompt, the physical environment, and the mathematical world brought forth to orient the path of exploration and allow the curricular path to drift in response to these interactions.

A teacher may be involved in setting boundaries for the lived curriculum, but if he or she attempts to direct its path towards particular topics or ideas, or attempts to force the students to reason about them in certain ways within those boundaries, the interaction ceases, at least for that moment, to be a conversation. A conversation is not conducted internally by any one person or object. This does not imply that educators have to abandon teaching the cultural tools of mathematical practice, but perhaps these tools can be shared with students in the course of conversation to broaden their understanding of the current issue of concern. Or perhaps a mathematical tool itself can be raised as a potential issue of concern. For example, standardized algorithms become an interesting topic rich with history when examined alongside invented strategies used for the same computation. In this way the teacher becomes involved in the conversation, not by imposing order, but by *sharing* his or her *understanding* and experience as a member of an extended mathematical community in order to broaden the present community's understanding of a mutual topic of concern. Similarly, students cannot be instructed to formalize as though it was a natural and necessary progression from concrete activities. Using mathematical formalizations, abstractions and generalizations must arise from the concerns addressed in the moment. A teacher may be able to raise concerns that are potentially explained through abstract forms of reasoning and may offer formalized explanations in response to the present concern. However, the concern raised may or may not become a mutual concern, and to be accepted, the explanation offered must be perceived by the students as broadening their understanding of the concern and allowing them to continue on.

Recognizing that the path of mathematical activity is contingent upon so many aspects of interaction is significant for creating an environment conducive to mathematical conversation. For example, choices made for the initial prompt, seating arrangements, access to rulers, graph paper, calculators, manipulatives, or any other object need to be considered as potential aspects of the common language created and the lived curriculum. Simply providing an appropriate physical environment, however, does not necessarily create a suitable space for conversation for all students. Students can only be invited into the space, they cannot be forced into participating conversationally within it.[3] The prepared aspects of the environment need to be considered along with the emotional basis of the interpersonal and mathematical interactions and the perceived expectations for mathematical activity.

Expectations for acceptable explanations present another concern for including mathematical conversations in the classroom. Inclusion of mathematical conversations necessitates an alteration in the present criteria for accepting mathematical explanations. Currently, mathematics education reforms are attempting to parallel mathematics as a discipline by defining the criterion for acceptance of mathematical explanations as the presentation of a "convincing argument" (Hersh 1993; NCTM 1989, 1991, 2000). A convincing argument is perceived as mathematically intelligent; however, this perception defines intelligence and acceptable mathematical practice based on an assumption that certainty is obtainable—at least for the immediate moment. This assumption is at odds with the emotional predisposition and the anticipated responses that occur in explanations in action offered within a practice of lingering. In chapter 7, adequate and intelligent mathematical action was attributed to persons with an ongoing emotional desire to *maintain one's relationships* and continue to *seek an understanding* of the experiences one has within the mathematical world of significance brought forth. This description of intelligence, if used as criteria for accepting mathematical explanations in the mathematics classroom, allows explanations offered within a mathematical conversation to be accepted as mathematical. Without such a change, explanations in action occurring within a mathematical conversation will consistently be judged as nonmathematical, as they are viewed as only precursors to "real" mathematics.[4]

Explanations as re-presentations, generally evoked when students are asked to summarize their explanations, are viewed as potentially valuable in both argumentation and conversation genres. In conversation, when emotional predispositions of mutual acceptance and curiosity are maintained, explanations as re-presentations become a potential vehicle for sharing and extending seemingly autonomous mathematical conversations in a classroom, creating a cultural network of conversations (Maturana 1988).

The lived curriculum as it occurs within a mathematical conversation potentially presents the strongest criticism against practically implementing mathematical conversations in the classroom. The teacher must accept that he or she is incapable of directing the conversation and must also be willing to be "addressed" or "moved" by the mathematics brought forth by students. In a community of mutual acceptance, the teacher may raise concerns arising out of the actions and interactions of students and must find a means for sharing, rather than imposing, the cultural tools of mathematics so they do not become "deadly" to the conversation. Both the teacher's role in the classroom and expectations surrounding the students' mathematical actions in a conversation must be supported by a curriculum that is viewed as a "domain for conversational action" (Applebee 1994). What is needed is not a revision of the current curriculum, but a re-vision of the mathematics curriculum. What might a mathematics curriculum that is viewed as a domain for conversational action look like?

Mathematical Conversations in the Classroom

Mathematical conversation as a gestural genre as presented here does not attempt to prescribe or even describe how mathematical conversations should be implemented into the classroom by defining the teacher's and students' roles in discourse, the tasks, or the learning environment (NCTM 1991). This work does, however, point to features of mathematical conversations and suggest ways in which they can be observed within the classroom. Mathematical conversation also points to the potential consequences for the choices educators make in terms of discourse and acceptable mathematical practices.

There are, of course, many questions that need to be addressed in relation to implementing mathematical conversations into the classroom. The primary questions suggested herein were:

- How is it possible to establish and maintain a mathematical community based on mutual acceptance whereby students feel responsible for attempting to understand the explanations of other students?
- How might an emotional predisposition of curiosity be fostered in the classroom to support and sustain mathematical relationships?
- What re-vision of the mathematics curriculum is necessary to allow the lived curriculum to be viewed as a domain for conversational action?

These questions do not have to be answered prior to attempting to implement mathematical conversations into the classroom, for providing recurrent opportunities for mathematical conversations may assist in establishing the emotional predispositions, expectations, and assumptions necessary for acting within and living out a conversational discourse and practice within mathematics.

The present research raises a number of pedagogical concerns for choosing argumentation as an exclusive form of discourse and problem solving, as it is currently perceived, as an exclusive form of practice. I offer mathematical conversations, not as a replacement, but as an alternative. While argumentation was frequently used as a contrast to mathematical conversation, these two gestural genres are not mutually exclusive. I also do not suggest that every interaction in mathematics should be an occasion for mathematical conversation. Mathematical conversation is one potential and possible path within the practice of mathematics requiring further exploration and explanation within mathematics education. What I offer, I am sure, is not the ultimate word on this issue of concern, for as Bakhtin (1984) has stated, "Nothing conclusive has yet taken place in the world, the ultimate word of the world and about the world has not yet been spoken, the world is open and free, everything is still in the future and will always be in the future" (166).

Re-turning to Features of Mathematical Conversation 149

Orienting Gestural Domains	Features within a Mathematical Conversation
ADDRESSIVITY TOWARDS THE OTHER *What response is expected from and offered to the other?* *What emotional predisposition is chosen?* *How are conflicts dealt with?*	*Oriented towards relationships among persons in interaction.* ♦ The listener has an ethical responsibility to respond to and address the other through questions, clarifications, or continuations and extensions of previous gestures. ♦ The mutual acceptance of self and of the other supports ongoing interpersonal relations. ♦ Alternative explanations and counterexamples are offered and perceived to assist in expanding understanding.
ADDRESSIVITY TOWARDS OTHERNESS *What response is expected from and offered by the participants and by the mathematical otherness?* *What emotional predisposition is chosen?*	*Oriented towards relationships among persons and the mathematical environment.* ♦ Participants seek insight and understanding by responding to and addressing otherness through questions, predictions, conjectures, and explanations. ♦ Otherness responds to the participants' gestures by bringing forth alternative and shifting mathematical landscapes. ♦ Curiosity and a structure of play support ongoing mathematical relations.
THE LIVED CURRICULUM Directing the Course *Who or what is responsible for directing the path of mathematical activity?* *What mechanisms set the boundaries of the lived curriculum?* Mathematical Explanations *How are mathematical explanations posed?* *What criteria are used to accept explanations?*	*Interactions between persons and mathematical environment lay a path of mathematical activity.* ♦ The direction of activity is led by no one and no thing. The path is oriented by emotionally charged mutual concerns arising in the moment. ♦ Proscriptive logic sets the boundaries: whatever is allowed by the mathematical situation and by the participants' experiences is acceptable. ♦ *Explanations in action* arise in the course of ongoing interaction and are open to revision at any time. ♦ *Explanations as re-presentation* arise in moments requiring a summary of interactions. ♦ Criteria of acceptance: When explanations are perceived to broaden understanding in that moment and are plausible, coherent with previous experiences and good enough for now; they allow participants to continue on. ♦ Explanations are observed to be accepted when they become part of the participants' subsequent actions.

Figure 27. Features of mathematical conversation[5]

Notes

1. It is probably better to say that the *expectations* surrounding the interaction allowed most participants to perceive the mathematical "initiate" as a prompt rather than a problem. We saw in chapter 6 how Ken and Stacey perceived the Arithmagon initially as a finite problem. Appendix B also suggests that the participants involved perceived the prompt, also the Arithmagon, as a finite problem.

2. The three examples of mathematical conversations in this work can be reread for demonstrating an ongoing emotional curiosity, gestures responding to the body of mathematics, and a recursive cycle of experiences, mutual concerns, and accepted explanations. These aspects of addressivity towards otherness allowed the conversants to maintain their relationships to the mathematics.

3. See Appendix B for an instance of interactions not consistent with mathematical conversation.

4. Ethnomathematics has suffered a similar fate. Using prefixes such as "ethno" or "proto" identify these forms of mathematics as "pre" mathematics or "cultural" mathematics but not "real" mathematics.

5. Figure 27 provides a summary of conversational features. The creation of the chart was a recursive and interpretative process involving reflection on the text, particularly chapters 5–7, and reflection in the process of writing of chapter 8. It did not exist prior to the writing of the body of the text, nor prior to the writing of the final chapter. However, it is not a product of these chapters either, as the summary of features created opportunities to rethink and revise both the body of the book and the final chapter. Its creation is another example of the "all at once" quality of the research (also see Appendix A).

Appendix A

The Research

Data for over fifteen pairs of participants have been collected for this study.[1] The participants include students in classroom and clinical situations from grades 3 to grade 10, undergraduate students in education, and parents and children involved in an extracurricular mathematics program. The number of mathematical sessions that each pair participated in varied from one to six. All sessions followed the same general format: a mathematical prompt was posed and it was expected that participants would continue to engage in it for as long as it held their interest. There was an expectation that this would be approximately one hour. The prompts were frequently found in problem-solving books (e.g., Mason, Burton, and Stacey 1985; Stevenson 1992) and made use of concrete manipulatives, paper and pencil, and computer software. The prompts were potentially rich in mathematics and allowed students with a variety of skills and experiences to participate in various ways. The mathematical prompts provided a starting point for inquiry into mathematical activity. While some outcomes were anticipated, it was not expected that all students would engage in similar activities. The pairs engaged in the mathematical environment in ways they determined appropriate and personally relevant.

Chapters 5, 6 and 7 present the mathematical activity of three different pairs of participants. Each chapter illustrates the activities of one pair of participants and discusses and interprets aspects of mathematical conversation in relation to their activity. While the interpretations made are directly related to the context of the illustrative example provided in each chapter, these interpretations were informed by observations made of all participants in this study—including sessions that were viewed as having qualities that were not consistent with mathematical conversations (see Appendix B for an example). The sessions selected for this monograph provide a broad basis from which to discuss the interdependent features of mathematical conversations; however, in relation to the theoretical underpinnings of this work (see chapter 1), it is inappropriate to say that all or even most of the features of conversation have been highlighted.

Chapter 5

Tamera and Kylie were originally contacted to participate in the Mathematical Understanding Project under Dr. Tom Kieren at the University of Alberta through a methods course for mathematics majors in secondary education. In this study Tamera and Kylie participated with other partners, but in similar sessions. After that study was completed they agreed to work with me on four additional sessions. The data included for this chapter arise from their activities in the fourth session.

An early version of this paper was presented at the American Educational Research Association (AERA) Annual Meeting in April, 1998 (see Gordon Calvert 1998). While the subject matter and Tamera and Kylie's work as described in chapter 5 is similar to the paper presented at AERA, the data re-presented here and the interpretations made have been revised considerably in the writing and rewriting of this manuscript.

Chapter 6

Stacey and Ken were asked by two researchers, Dr. Tom Kieren and David Reid, at that time a graduate student working with Dr. Kieren, to participate in the Mathematical Understanding Project. These two researchers collected data such as videotapes, artifacts of Stacey and Ken's work, as well as field notes. Another researcher, Elaine Simmt, and I were asked to become part of the research community for interpreting Stacey and Ken's actions and interactions from within an enactivist framework. This community was a unique opportunity for research, as each of us interacted with the same data but brought different perspectives to it, such as emphases on mathematical understanding, reasoning, beliefs, and conversation. Although these perspectives were different, the data and our commitment to the enactivist community provided an occasion in which we could share interpretations. This data along with the different interpretive perspectives were also presented at several conferences; therefore, the questions and interpretations of the audience also have been folded into the interpretations stated herein. The interpretations throughout this work then, cannot be said to be solely my own, but "half someone else's"(Bakhtin 1981, 293). I do not know where my interpretations end and my colleagues' begin. Yet, this also becomes a topic for this chapter, as a conversation is not a patchwork of individual products but an overlapping of thoughts, ideas, questions, and stories. Needless to say, I am grateful for the use of this data for chapter 6.

Stacey and Ken's work was the focus of a paper presented at the AERA Annual Meeting in April 1995 (see Kieren et al. 1995). While this earlier paper informed the writing of chapter 6, the recursive process of revisiting the data and rewriting interpretations in light of later research has substantially altered the content of the original paper.

Chapter 7

Jolene and her son Calvin were registered participants in a ten-week course conducted by Professor Elaine Simmt from the University of Alberta. Six parent-child pairs met weekly to interact with one another and a mathematical prompt. Both children and parents brought a diverse set of skills, experiences, and abilities. The intent of the program was for children and their parents to enrich their understanding of mathematics. Because child and parent worked together, not as tutor-tutee, new relationships were fostered between parent and child, and between the participants and mathematics as a whole.

The narrative, including dialogue, actions, and drawings, was reconstructed from field notes taken during and after the session. Field notes included sketches of Calvin and Jolene's work, parts of dialogue, activities, and interpretations of these events. These notes were exchanged with Professor Simmt, who provided further information and interpretations of events.

Jolene and Calvin's work was the data used in subject matter for papers presented at the AERA Annual Meeting in March 1997 (see Gordon Calvert 1997) and in April 1999 (see Gordon Calvert 1999). Both papers informed the interpretations made throughout chapter 7.

Chapter 8

This final chapter summarizes many of the features of mathematical conversation under three headings: Addressivity towards the other; addressivity towards otherness; and the lived curriculum. These "gestural domains" were developed by returning to data from chapters 5, 6, and 7 and drawing from these chapters generalized statements referencing the features of mathematical conversation as they occurred in context. Common themes and relationships were noted between statements, and they were grouped together to develop a number of categories. Eventually the categories were refined to the three domains noted above. These gestural domains, along with their significant features, were organized into Figure 27 (page 149). Chapter 8 was written by referencing this chart; however, the chart was also revised as the writing of this chapter occurred. Similarly, aspects of chapters 5, 6, and 7 were also revised and clarified in this recursive process.

Note

1 All names have been changed to protect the participants' anonymity.

Appendix B

Interactions Not Consistent with Features of Mathematical Conversation

Not every student at all times is capable of or willing to engage in a mathematical conversation. Rather than speaking in generalizations, the following example illustrates an interaction that contains elements that are not consistent with features of a conversation.

While observing Melinda and Randy,[1] a pair of undergraduate students working on the Arithmagon, I noticed that their body language kept them distant and their verbal exchanges were often terse and occasionally sarcastic. Their writing about their experiences provided additional insight into their thoughts and emotions in this setting. Melinda wrote:

> I found that it might have been easier to do the problem myself because I wouldn't have spent as much time on the problem. . . . Randy was trying to overanalyze the problem which influenced me to do so. . . . It helped when Dave [another researcher present] asked us to just write down everything we knew about B & C. He should have done so sooner. . . . This, however, did confirm that we were on the right track and did figure out the problem correctly. We spent too much time trying to express ourselves or rather to express the generalizations into sentences. We should have had the answer a lot sooner.

Melinda comments that the problem shouldn't be overanalyzed, it had a right track, a correct answer, and that the answer should come quickly. These expectations indicate that she views the purpose for engaging in the activity as finding an answer to a well defined pre-given problem as quickly and as efficiently as possible. Once she found an answer to what she thought was the problem her relationship to the mathematics ended. There was no desire to explore it further. It is not surprising that she rejected the usefulness of interacting with a peer, but appreciated the help from an authority. Trying to express one's thinking would have simply gotten in the way of what she saw as the goal of this session.

Randy, on the other hand, was not able to engage in a conversation for another reason. He wrote:

> A lot of my difficulties stemmed from the fact that I was highly embarrassed that I did not really know how to do the exercise and yet had to try. I did not want to dominate the conversation and felt like giving way all the time to her point of view.

Randy's writing points to the quality of addressivity. That is, he did not feel he existed in a relationship of mutual acceptance. His inability to find a way to do the problem heightened his anxiety and self consciousness. Randy continued to express his discomfort even after his third session. He wrote, "Was this purposely supposed to be intimidating or was it just my own mind set at the time?" Even though the same task and research setting was used successfully with a number of other pairs who engaged in conversation, this example illustrates that the ability or willingness to enter into a conversation based on a relationship of mutual acceptance is not automatic upon providing a suitable environment.

Note

1   Names have been changed to protect anonymity.

References

Abram, D. (1996). *The spell of the sensuous: Perception and language in a more-than-human world.* New York: Pantheon Books.

Applebee, A. N. (1994). Toward thoughtful curriculum: Fostering discipline-based conversation. *English Journal* 83(3): 45–51.

Atkins, S. L. (1999). Listening to students: The power of mathematical conversations. *Teaching Children Mathematics,* 5: 289–295.

Azmitia, M. (1990). Peer interactive minds: Developmental, theoretical, and methodological issues. In P. B. Baltes, and U. M. Staudinger, eds., *Interactive minds: Life-span perspectives on the social foundation of cognition,* 133–162. Cambridge: Cambridge University Press.

Bakhtin, M. M. (1981). *The dialogic imagination,* M. Holquist, ed. and trans., C. Emerson, trans.. Austin: University of Texas Press.

———. (1984). Problems of Dostoevsky's poetics. In C. Emerson, ed. and trans., *Theory and history of literature,* volume 8, Minneapolis: University of Minnesota Press.

———. (1986). *Speech genres and other late essays.* C. Emerson and M. Holquist, eds., V. W. McGee, trans.. Austin, TX: University of Texas Press.

Ball, S. J. (1990). *Politics and policy making in education: Explorations in policy sociology.* London: Routledge.

Barrow, J. D. (1992). *Pi in the sky: Counting, thinking and being.* London: Penguin.

Bateson, G. (1972). *Steps to an ecology of mind.* New York: Ballantine.

———. (1979). *Mind and nature: A necessary unity.* New York: Bantam.

Bateson, M. C. (1989). *Composing a life.* New York: Plume.

Belenky, M. F., Clinchy, B. M., Goldberger, N. R., and Tarule, J. M. (1986). *Women's ways of knowing: The development of self, voice and mind.* New York: Basic.

Berkowitz, M. W., Oser, F., and Althof, W. (1987). The development of sociomoral discourse. In M. Kurtines, and J. L. Gewirtz, eds., *Moral development through social interaction,* 322–352. New York: Wiley.

Billig, M. (1987). *Arguing and thinking: A rhetorical approach to social psychology.* Cambridge: Cambridge University Press.

Brown, T. (1994). Creating and knowing mathematics through language and experience. *Educational Studies in Mathematics* 27: 79–100.

Bruner, J. (1986). *Actual minds, possible worlds.* Cambridge, MA: Harvard University Press.

Carse, J. P. (1986). *Finite and infinite games: A vision of life as play and possibility.* New York: Ballantine.

Charles, R. I., and Lester, F. K. (1982). *Teaching problem solving: What, why and how.* Palo Alto, CA: Dale Seymour.

Clinchy, B. M. (1996). Connected and separate knowing: Toward a marriage of two minds. In Goldberger, N., Tarule, J., Clinchy, B., and Belenky, M., eds., *Knowledge, difference, and power: Essays inspired by women's ways of knowing,* 205–247. New York: Basic Books.

Cobb, P. (1994). Where is the mind? Constructivist and sociocultural perspectives on mathematical development. *Educational Researcher* 23(7): 13–20.

Cobb, P., Boufi, A., McClain, K., and Whitenack, J. (1997). Reflective discourse and collective reflection. *Journal for Research in Mathematics Education* 28, 258–277.

Cobb, P., Yackel, E., and Wood, T. (1992). Interaction and learning in mathematics classroom situations. *Educational Studies in Mathematics* 23: 99–122.

———. (1995). The teaching experiment classroom. In P. Cobb, and H. Bauersfeld, eds., *The emergence of mathematical meaning*, 17–24. Hillsdale, NJ: Lawrence Erlbaum Associates.

Code, L. (1991). *What can she know? Feminist theory and the construction of knowledge*. Ithaca, NY: Cornell University Press.

Davis, B. (1994). Listening to reason: An inquiry into mathematics teaching. Unpublished doctoral dissertation. University of Alberta, Edmonton, Alberta.

———. (1995a, April). Why I'm a cat person (Chair's response). T. Kieren, L. Gordon Calvert, D. Reid and E. Simmt, Coemergence: Four enactive portraits of mathematical activity. Paper presented at the annual meeting of the American Educational Research Association, San Francisco, California.

———. (1995b). Why teach mathematics? Mathematics education and enactivist theory. *For the Learning of Mathematics* 15(2): 2–9.

Davis, P. J., and Hersh, R. (1981). *The mathematical experience*. Boston, MA: Houghton Mifflin.

———. (1986). *Descartes' dream: The world according to mathematics*. Boston, MA: Harcourt Brace Jovanovich.

Eco, U. (1994). *Six walks in the fictional woods*. Cambridge: Harvard University Press.

Elbow, P. (1973). *Writing without teachers*. London: Oxford University Press.

———. (1986). *Embracing contraries*. New York: Oxford University Press.

Ellsworth, E. (1989). Why doesn't this feel empowering? Working through the repressive myths of critical pedagogy. *Harvard Educational Review* 59: 297–324.

Epstein, D., and Levy, S. (1995). Experimentation and proof in mathematics. *Notices of the American Mathematical Society* 42: 670–674.

Epstein, D., Levy, S., and de la Llave, R. (1996). Statement of philosophy and publishability criteria. *Experimental Mathematics homepage*. [www.expmath.com/expmath/philosophy.html].

Ernest, P. (1994). The dialogical nature of mathematics. In P. Ernest, ed., *Mathematics, education and philosophy*, 33–48. London: Falmer Press.

Foucault, M. (1980). In C. Gordon, ed., *Power/knowledge: Selected interviews and other writings, 1972–1977*. New York: Pantheon Books.

Gadamer, H. G. (1989). *Truth and method*, 2nd rev. ed., J. Weinsheimer and D. G. Marshall, trans.. New York: Continuum.

Goldberger, N., Tarule, J., Clinchy, B., and Belenky, M., eds. (1996). *Knowledge, difference, and power*. New York: Basic Books.

Goldenberg, E. P. (1991). Mathematical conversation with fourth graders. *Arithmetic Teacher* 38(8): 38–43.

Goodman, N. D. (1993). Modernizing the philosophy of mathematics. In A. White, ed., *Essays in humanistic mathematics*, 63–66. Washington, DC: Mathematical Association of America.

Gordon Calvert, L. (1997, March). Lingering in a mathematical space: An alternative view to problem solving. Paper presented at the Annual Meeting of the American Educational Research Association, Chicago, IL.

———. (1998, April). Mathematical conversations: Explanations in the process of understanding. Paper presented at the Annual Meeting of the American Educational Research Association, San Diego, CA.

———. (1999, April). The nature of mathematics practice: Is mathematics problem solving? Paper presented at the Annual Meeting of the American Educational Research Association, Montreal, Quebec.

Grondin, J. (1995). *Sources of hermeneutics*. Albany, NY: State University of New York.

Grudin, R. (1990). *The grace of great things*. New York: Ticknor and Fields.

Hadamard, J. (1973). *The mathematician's mind: The psychology of invention in the mathematical field*, rev. ed.. NJ: Princeton University Press. (Originally published in 1945).

Haroutunian-Gordon, S., and Tartakoff, D. S. (1996). On the learning of mathematics through conversation. *For the Learning of Mathematics* 16(2): 2–10.

Hersh, R. (1993). Proving is convincing and explaining. *Educational Studies in Mathematics* 24: 389–399.

Hicks, D. (1996a). *Discourse, learning, and schooling.* Cambridge: Cambridge University Press.

———. (1996b). Discourse, learning, and teaching. *Review of Research in Education* 21: 49–95.

Horgan, J. (1993). The death of proof. *Scientific American* 269(4): 92–103.

Jardine, D. W., and Field, J. C. (1996). Restoring [the] life [of language] to its original difficulty: On hermeneutics, whole language, and "authenticity." *Language Arts* 73: 261–270.

Johnson, M. (1987). *The body in the mind: The bodily basis of meaning imagination, and reason.* Chicago: University of Chicago Press.

Kieren, T., Gordon Calvert, L., Reid, D., and Simmt, E. (1995, April). Coemergence: Four enactive portraits of mathematical activity. Paper presented at the Annual Meeting of the American Educational Research Association, San Francisco, California.

Kleiner, I., and Movshovitz-Hadar, N. (1997). Proof: A many-splendored thing. *The Mathematical Intelligencer* 19(3): 16–26.

Krummheuer, G. (1995). The ethnography of argumentation. In P. Cobb, and H. Bauersfeld, eds., *The emergence of mathematical meaning: Interaction in the classroom cultures*, 229–269. Hillsdale, NJ: Lawrence Erlbaum Associates.

Lakatos, I. (1976). *Proofs and refutations: The logic of mathematical discovery.* Cambridge: Cambridge University Press.

———. (1986). A renaissance of empiricism in the recent philosophy of mathematics? In T. Tymoczko, ed., *New directions in the philosophy of mathematics: An anthology*, 29–48. Boston: Birkhäuser.

Lakoff, G., and Johnson, M. (1980). *Metaphors we live by.* Chicago: University of Chicago Press.

Lakoff, G., and Núñez, R. E. (1997). The metaphorical structure of mathematics: Sketching out cognitive foundations for a mind-based mathematics. In L. D. English, ed., *Mathematical reasoning: Analogies, metaphors, and images*, 21–89. Mahwah, NJ: Lawrence Erlbaum Associates.

Lampert, M. (1990). When the problem is not the question and the solution is not the answer: Mathematical knowing and teaching. *American Educational Research Journal* 27(1): 29–63.

Lampert, M., Rittenhouse, P., and Crumbaugh. (1996). Agreeing to disagree: Developing sociable mathematical discourse. In D. R. Olson, and N. Torrance, eds., *The handbook of education and human development: New models of learning, teaching and schooling*, 731–764. Cambridge, MA: Blackwell.

Lester, F. K. (1994). Musings about mathematical problem-solving research: 1970–1994. *Journal for Research in Mathematics Education* 25: 660–675.

Lewis, M., and Simon, R. I. (1986). A discourse not intended for her: Learning and teaching within patriarchy. *Harvard Educational Review* 56: 457–472.

Luce-Kapler, R. (Chair). (1996, June). Buttons: A response to Elaine and Lynn (Chair's response). T. Kieren, L. Gordon Calvert, and E. Simmt, Interaction and mathematics knowing: An interactive panel. Paper presented at the Annual Meeting of the Canadian Society for the Study of Education, Brock University, St. Catharines, Ontario.

Mason, J., Burton, L., and Stacey, K. (1985). *Thinking mathematically*, rev. ed.. Wokingham, England: Addison-Wesley.

Maturana, H.R. (1985). Interview with Humberto Maturana. *Associazione Oikos home page*. [www.oikos.org/maten.htm].

———. (1987). Everything is said by an observer. In W. I. Thompson, ed., *Gaia a way of knowing: Political implications of the new biology*, 65–82. Hudson, NY: Lindisfarne Press.

———. (1988). Reality: The search for objectivity or the quest for a compelling argument. *Irish Journal of Psychology* 9: 25–82.

———. (1991). Science and daily life: The ontology of scientific explanations. In F. Steier, ed., *Research and reflexivity*, 30–52. London: Sage Publications.

———. (1998, October). The biology of cognition: Languaging, emotioning and reality. A presentation given at the University of Calgary. Calgary, Alberta.

Maturana, H., and Varela, F. (1980). *Autopoesis and cognition: The realization of the living.* Boston: D. Reidel.

———. (1987). *The tree of knowledge: The biological roots of human understanding,* rev. ed. Boston: Shambhala Publications.

Minsky, M. (1985). *The society of mind.* New York: Simon and Schuster.

Morson, G. S., ed. (1981). *Bakhtin: Essays and dialogues on his work.* Chicago: University of Chicago Press.

Nachmanovitch, S. (1990). *Free play: The power of improvisation in life and the arts.* New York: Putnam's.

NCTM. (1980). *An agenda for action: Recommendations for school mathematics of the 1980s.* Reston, VA: National Council of Teachers of Mathematics.

———. (1989). *Curriculum and evaluation standards for school mathematics.* Reston, VA: National Council of Teachers of Mathematics.

———. (1991). *Professional standards for teaching mathematics.* Reston, VA: National Council of Teachers of Mathematics.

———. (2000). *Principles and standards for school mathematics.* Reston, VA: National Council of Teachers of Mathematics.

Noddings, N. (1994). Does everybody count? Reflections on reforms in school mathematics. *Journal of Mathematical Behaviour* 13: 89–104.

Oakeshott, M. (1989). *The voice of liberal learning: Michael Oakeshott on Education,* T. Fuller, ed.. New Haven: Yale University Press.

Oxford English Dictionary, 2nd ed. [Electronic Version]. (1992). Oxford: Oxford University Press.

Pinar, W. F., and Grumet, M. R. (1976). *Toward a poor curriculum.* Dubuque, IA: Kendall/Hunt.

Polya, G. (1954). *Mathematics and plausible reasoning: Induction and analogy in mathematics.* Princeton, NJ: Princeton University Press.

———. (1973). *How to solve it: A new aspect of mathematical method,* 2nd ed. Princeton, NJ: Princeton University Press.

Popper, K. R. (1945). *The open society and its enemies.* London: Routledge.

Putnam, R. T., Lampert, M., and Peterson, P. L. (1990). Alternative perspectives of knowing mathematics in elementary schools. *Review of Research in Education* 16: 57–150.

Ruddick, S. (1996). Reason's "femininity": A case for connected knowing. In N. Goldberger, J. Tarule, B. Clinchy, and M. Belenky, eds., *Knowledge, difference, and power: Essays inspired by women's ways of knowing,* 248–273. New York: Basic Books.

Sawada, D. (1991). Deconstructing reflection. *The Alberta Journal of Educational Research* 37: 349–366.

Sfard, A., Nesher, P., Streefland, L., Cobb, P., and Mason, J. (1998). Learning mathematics through conversation: Is it as good as they say? *For the Learning of Mathematics* 18(1): 41–51.

Shotter, J. (1993). *Conversational realities: Constructing life through language.* London: Sage.

———. (1995). In conversation: Joint action, shared intentionality and ethics. *Theory and Psychology* 5: 49–73.

Simpson, D. (1995). *The academic postmodern and the rule of literature: A report on half-knowledge.* Chicago: University of Chicago Press.

Sinclair, J. M. and Coulthard, R. M. (1975). *Towards an analysis of discourse: The English used by teachers and pupils.* London: Oxford University Press.

Smith, D. G. (1991). Hermeneutic inquiry: The hermeneutic imagination and the pedagogic text. In E. C. Short, ed., *Forms of curriculum inquiry,* 187–209. Albany: State University of New York Press.

Spolin, V. (1963). *Improvisation for the theatre.* Evanston, IL: Northwestern University Press.

Steier, F. (1995). From universing to conversing: An ecological constructionist approach to learning and multiple description. In L. P. Steffe, and J. Gale, eds., *Constructivism in education,* 67–84. Hillsdale, NJ: Lawrence Erlbaum Associates.

Stevenson, F. W. (1992). *Exploratory problems in mathematics.* Reston, VA: National Council of Teachers of Mathematics.

Sumara, D. J. (1996). *Private readings in public: Schooling the literary imagination.* New York: Peter Lang.

Taylor, C. (1991). The dialogical self. In D. R. Hiley, J. F. Bohman, and R. Shusterman, eds., *The interpretive turn: Philosophy, science, culture,* 304–314. Ithaca, NY: Cornell University Press.

Tomm, K. (1989, January). Humberto Maturana's view of cognition. A presentation given at the University of Alberta. Edmonton, Alberta.

Triadafillidis, T. A. (1998). Dominant epistemologies in mathematics education. *For the Learning of Mathematics* 18(2): 21–27.

Tymoczko, T., ed. (1986). *New directions in the philosophy of mathematics: An anthology.* Boston: Birkhäuser.

———. (1993). Humanistic and utilitarian aspects of mathematics. In A. White, ed., *Essays in humanistic mathematics,* 11–14. Washington, DC: Mathematical Association of America.

Ushers, R., and Edwards, R. (1994). *Postmodernism and education.* London: Routledge.

Varela, F. (1987). Laying down a path in walking. In W. I. Thompson, ed., *Gaia a way of knowing: Political implications of the new biology,* 48–64. Hudson, NY: Lindisfarne Press.

Varela, F. J., Thompson, E., and Rosch, E. (1991). *The embodied mind: Cognitive science and human experience.* Cambridge, MA: MIT Press.

von Glasersfeld, E. (1987). Learning as a constructive activity. In C. Janvier, ed., *Problems of representation in the teaching and learning of mathematics,* 3–17. Hillsdale, New Jersey: Lawrence Erlbaum Associates.

———. (1988). The reluctance to change a way of thinking. *Irish Journal of Psychology* 9: 83–90.

———. (1990). Distinguishing the observer: An attempt at interpreting Maturana. *Associazione Oikos home page.* [www.oikos.org/vonobserv.htm].

———. (1995a). A constructivist approach to teaching. In L. P. Steffe, and J. Gale, eds., *Constructivism in education,* 3–15. Hillsdale, NJ: Lawrence Erlbaum Associates.

———. (1995b). *Radical constructivism: A way of knowing and learning.* London: Falmer Press.

Whitin, D. J., and Wilde, S. (1995). *It's the story that counts: More children's books for mathematical learning K–6.* Portsmouth, NH: Heinemann.

Winograd, T., and Flores, F. (1986). *Understanding computers and cognition.* Reading, MA: Addison-Wesley.

Wittgenstein, L. (1953). *Philosophical investigations,* G. E. M. Anscombe, trans.. New York: Macmillan.

Yackel, E., and Cobb, P. (1996). Sociomathematical norms, argumentation, and autonomy in mathematics. *Journal for Research in Mathematics Education* 27: 458–477.

Yackel, E., Cobb, P., and Wood, T. (1991). Small-group interactions as a source of learning opportunities in second-grade mathematics. *Journal for Research in Mathematics Education* 22, 390–408.

Index

Abstraction; see Formalization
Addressivity 55–56, 127, 133, 144, 149, 153–154, 158
   towards the other 133–139
   towards otherness 139–142, 150
Algorithm 27, 93, 103–104, 142, 143, 145
All-at-once 5, 6–7, 56, 150
Argument
   is war 18, 20, 21, 43
Argumentative discourse 18, 134–135
   difficulties with 18–21
   and knowledge acquisition 21–22
   and mathematical practice 15–17, 46, 82, 83, 127, 148
   and problem solving 2, 8, 30, 37
Authority 2, 83

Boundary 10, 83, 96, 103, 126–129, 145

Classroom discourse
   and teaching math 2, 16, 138–139, 141–142, 146–147
Cognition 2, 51–52, 56
   and evolution 123–124
   and problem solving 125
Cognitivism 28–29, 30, 38, 48–49
Constructivism 38, 43, 48
   classroom roles 29–30, 46
   and problem solving 29–31, 140
Conversari 47
Convince
   etymology 43

Counterexamples 15, 16, 22, 30, 79–81, 127, 137–138
Critical rationalism 18
Culture 17, 39, 50, 52, 84, 147
   role playing 96–97
Curriculum 8, 133, 142–147, 149

Discourse of disagreement 17, 19–20, 46

Discurrere 2–3

Emotions
   and conflicts 20, 135–136, 146
   and doing mathematics 1, 20, 27, 40, 109–110, 131, 139–142, 146, 150, 157–158
   and games 125–129, 130
   and validating explanations 40–42, 82–84
Enactivism 2, 9, 48–53, 58–59
Errors 1, 32, 92, 127
Ethics
   and interaction 39, 55, 102, 137
Experimental methods
   in mathematics 33–35
Experiri 34
Explanations
   criterion for acceptance 9, 63–65, 72–73, 79–84, 146, 149
   in action 71–72, 81–83, 84, 144, 149
   validate 38, 64, 71, 80, 82, 102
   as re-presentations 81–83, 144, 146, 149

Feminist educators 20–21
Formalization
    and abstraction 72–75, 145

Game
    doubting 21
    finite and infinite 125–129
    and math activity 1, 10, 105
Gestures 54, 94, 96, 109–110, 133–136, 137, 139, 140, 143
Gestural genre 132–137, 139–140, 145, 147, 148
Goal
    -directed 47, 123, 124
    in interaction 18, 22, 94
    of learning mathematics 26, 27, 103
    of living 124
    metaphor 127–127
    of playing games 126, 128
    predetermined 9, 47, 57, 109, 123
    of problem solving 28, 29, 30, 93, 103, 105, 110, 121
Good enough 64, 65, 71, 124, 132, 144

Hermeneutics 9, 51, 53, 54

Improvisation 87–88, 94
Incompleteness 6–7, 54, 65, 119, 132
Intelligence 8, 30 40–41, 123–124, 125, 127; see also Cognition and Intelligent action
Intelligent action 10, 26, 27–31, 52, 123–124, 146
Interaction
    forms of 92–93
    us/not-us 50, 52, 53, 59
Inter-objectivity 39
Intimacy 94–97, 137–138
IRF pattern of discourse 28–29, 45

Knowing, Ways of
    connected/narrative 22–23, 32–33, 37
    logico-scientific or paradigmatic 22
    and reality 29
    separate 21–22

Western culture 25
Knowledge; see also Knowing, Ways of
    assumptions 23, 25
    mathematical 13–14, 27, 29, 32, 42
    objective 22

Language 6, 16, 50, 53–56, 96, 119, 145
Listener 83, 93, 97, 98, 134, 135
Love 96; see also Mutual acceptance

Mathematics 112
    body of 35–38, 42, 52, 72, 75, 93, 102, 116, 117, 119, 122, 131, 139, 141
    definition 32
    explanatory domain 83–84, 131
    history of 36–37, 51, 52, 116, 117, 130, 145, 146
    imaginary landscape 101–104, 106
    philosophy of 2, 8, 14, 15, 17, 31, 33
    real 84, 150
Monologue 134–135
Mutual acceptance 41, 80, 96, 102, 136–139, 144, 146–147, 158

Perturbation 29–30
Play 7, 9, 10, 83, 105
    and games 126–129
    and mathematical activity 104–106, 109, 141
    purpose of 105–106, 110
    -space 87, 141
Power 3, 18, 19, 20, 21, 40
Practice
    word usage 3, 10.
Prejudice 52–53, 63, 83, 84
Prompts 5, 58–59, 62–63, 65–66, 90, 92–94, 114, 130, 141, 145, 151
Proof 3, 13, 22, 34–35, 37, 52, 64; see also Explanations
    and mathematical practice 14–15, 31, 34, 127
Proofs and Refutations 15

Index

Random 3, 39, 47, 50, 75, 109, 123, 124
Reality
    conversarial 38–42, 49, 53, 56, 57, 64, 73, 80, 82, 83, 85, 102, 136–138
    multiversal 38–42, 80, 82
    universal 39–42, 56, 82, 85, 102, 139
Reasoning
    demonstrative and plausible 14–15, 31–32
Relationships
    maintaining 9, 93, 97, 98, 108, 110, 118, 122–124, 129, 141, 146, 157
    re-pair 97–98

Research methods 5, 11, 7–8, 151–154; see also All-at-once and Enactivism
Ritual 3, 94, 142

Speech genre 55, 84, 133

Talk
    in math 2, 16, 45, 46
Technology 8, 25–27

Utterances 54–55, 118, 119, 133

Viability 29, 64, 71–72

Studies in the Postmodern Theory of Education

General Editors
Joe L. Kincheloe & Shirley R. Steinberg

Counterpoints publishes the most compelling and imaginative books being written in education today. Grounded on the theoretical advances in criticalism, feminism, and postmodernism in the last two decades of the twentieth century, Counterpoints engages the meaning of these innovations in various forms of educational expression. Committed to the proposition that theoretical literature should be accessible to a variety of audiences, the series insists that its authors avoid esoteric and jargonistic languages that transform educational scholarship into an elite discourse for the initiated. Scholarly work matters only to the degree it affects consciousness and practice at multiple sites. Counterpoints' editorial policy is based on these principles and the ability of scholars to break new ground, to open new conversations, to go where educators have never gone before.

For additional information about this series or for the submission of manuscripts, please contact:
        Joe L. Kincheloe & Shirley R. Steinberg
        c/o Peter Lang Publishing, Inc.
        275 Seventh Avenue, 28th floor
        New York, New York 10001

To order other books in this series, please contact our Customer Service Department:
        (800) 770-LANG (within the U.S.)
        (212) 647-7706 (outside the U.S.)
        (212) 647-7707 FAX

Or browse online by series:
        www.peterlangusa.com